Research Reports in Physics

W0050951

Research Reports in Physics

H. Schmidt-Böcking A. Schempp
K. E. Stiebing (Eds.)

Materials Research with Ion Beams

With 82 Figures

Springer-Verlag Berlin Heidelberg GmbH

Editors

Prof. Dr. Horst Schmidt-Böcking
Dr. Kurt E. Stiebing
Institut für Kernphysik
Universität Frankfurt
August-Euler-Straße 6, W-6000 Frankfurt am Main 90

Priv. Doz. Dr. Alwin Schempp
Institut für Angewandte Physik
Universität Frankfurt
Robert-Mayer-Straße 2-4, W-6000 Frankfurt am Main 11

ISBN 978-3-540-55774-6   ISBN 978-3-662-02794-3 (eBook)
DOI 10.1007/978-3-662-02794-3

Typesetting: Camera ready by the author/editor
57/3140 - 5 4 3 2 1 0 - Printed on acid-free paper

Preface

It was the purpose of this symposium to provide information about the present status of selected topics from nuclear and applied physics as related to materials research with ions beams and to stimulate the interchange of experience and ideas. Such links between basic and applied research activities are particularly important at the present time where recent technological developments open new possibilities for experimental approaches, such as the production of more and more exotic ion beams (e.g. beams of slow very highly charged ions, beams of ion clusters, etc.). Here the interaction of the projectile with target atoms becomes increasingly complex as do the spectra of observables obtained from collisions of such ions with atoms or solids. Therefore, a detailed knowledge of the reaction mechanisms is crucial for an efficient use of such beams in materials science.

This symposium has been held on the occasion of the 60th birthday of Prof. Dr. Klaus Bethge, Institut für Kernphysik, and Prof. Dr. Horst Klein, Institut für Angewandte Physik der Johann Wolfgang-Goethe Universität, Frankfurt am Main. Very recently, activities of these two institutes have merged with the funding of an "ECR-RFQ" installation at the Institut für Kernphysik. This new ion beam facility will provide intense beams of highly charged ions in an energy range well suited for basic atomic research as well as for materials analysis and modification.

The recent progress in ion source and accelerator technology provides us with compact devices at comparatively low costs for producing sufficiently intense beams of highly charged ions with variable velocity. These new experimental possibilities supply very selective methods for both atomic physics and materials analysis and modification and will clearly stimulate a large number of activities in these fields in the near future. This has already been demonstrated by quite a number of research groups all over the world. For an overview of existing results see, e.g. the references cited in the article of the editors in this book.

By means of the extra ionization power carried in the atomic shells of highly ionized heavy ions in form of potential energy, the use of such highly charged ions offers opportunities not afforded by traditional methods. In contrast to this, singly or doubly charged heavy ions as well as light ions such as protons and alpha particles cause ionization only by means of their kinetic energy (dynamic ionization), a degree of freedom, which, of course, is also present when highly charged ions are used. Furthermore, atomic collision physics has proven that a highly charged ion, dependent on its actual charge state (or even metastable state of excitation) and on its velocity, may induce very selective reactions. These reactions can no longer be described in terms of mean charge states or mean values of the stopping power. Charge transfer into selected states and Auger- and photon cascades when the ion penetrates the surface give rise to very structured electron or photon spectra containing detailed information on static and dynamical properties of electron distributions in the surface region. Therefore, the use of such beams opens new frontiers in materials research, provided a detailed knowledge of the ion-atom/surface reaction mechanisms can be supplied by basic atomic-physics research.

An additional attractive property of slow highly charged ions is that they will release their potential energy of several keV in a very short distance in the surface layer, ionizing all neighbouring atoms in the impact region. Therefore, well known shortcomings of SIMS may be overcome by using slow highly charged ions. Basically new techniques of surface analysis and modification are feasible such as the use of pulsed μ-beams of highly charged ions for controlled element analysis of selected surface regions in the μm range.

This book is intended to give a partial review of progress in the field of atomic physics with slow and fast highly charged ions. Recent results on charge exchange in and near surfaces as well as charge exchange during penetration of single crystals under channeling conditions are described. The use of channeling to obtain detailed information on the structure of complex crystalline compounds such as, for instance, high-temperature superconductors is discussed. Special high resolution techniques of nuclear physics and their application in materials re-search and the use of high energy heavy ion beams for materials modification are presented. The importance and physical problems of the ion-wall interaction in fusion reactors is discussed. New possible applications of cluster beams in mate-rials modification and the application of ion implantation to the development of new solid state detectors are presented. Finally, the new ECR-RGQ installation at the Institut für Kernphysik in Frankfurt is described and an outline of the planned research program at this facility is given.

The editors and organizers of the symposium express their gratitude to all authors and to those who helped during the organization of the symposium and the preparation of this book, in particular H. Latka, Ch. Thimmel, M. Stiebing, and R. Thomae. The financial support for the organization of the symposium by industry (NTG) and the Bundesministerium für Forschung und Technologie (BMFT) is gratefully acknowledged.

Frankfurt am Main
March 1992

H. Schmidt-Böcking
A. Schempp
K.E. Stiebing

Contents

Ion Tracks in Materials Research and Microtechnology

R. Spohr

Gesellschaft für Schwerionenforschung mbH,
Postfach 110 552, W-6100 Darmstadt, Fed. Rep. of Germany

The phenomenon of discrete, directed, and developable ion tracks opens the possibility to structure and modify solids with individual atomic particles on a mesoscopic scale at a high level of predetermination. The depth and sharpness of this microtool foreshadows new stimuli for a variety of fields.

The technique is based on the possibility to chemically amplify a quasi one-dimensional damage zone — engraved by the passage of an ion through an insulating solid — by several orders of magnitude, a key feature in practical applications.

The preliminary exploration of the technique and its relevant properties for materials research and microtechnology are lined out.

- **Basic features of ion tracks**
  - creation of defects
  - enhancement and annealing of defects
  - selective chemical etching of latent ion tracks
  - suitable parameters for influencing the results

- **Ion track applications**
  - single pores in medicine and low temperature physics
  - multiple pores in filtration and membrane technology
  - creation of directed (anisotropic) properties in solids
  - prospects of heavy-ion lithography
  - low energy ion tracks for generating surface texture
  - focussed ion beams for creating predetermined shapes

Introduction

Lithography — based on the precisely controlled spreading of etch fronts — is now gradually leaving the realm of planar technology and penetrating into the third dimension. Besides electronic properties this opens the access to mechanical, optical, magnetic, thermal, and chemical properties requiring often significant depths. The resulting sensors and actors are combined to microsystems capable to communicate with each other on a higher level. This combination of microequipment requires deep-cutting lithographic tools with high lateral control and depth-resolution. Ion tracks represent such a tool in which the initial conditions are defined by a one-dimensional engravement ruling the progression of the development process.

Historically, track technology has its earliest roots in the cloud chamber of Wilson which in 1898 enabled the observation of individual alpha particles in supersaturated moist air [1]. Thereby a string of microscopic water droplets is formed, triggered by secondary ions created along the path of an alpha particle. The rapidly fading phenomenon revealed a direct glimpse into the atomic

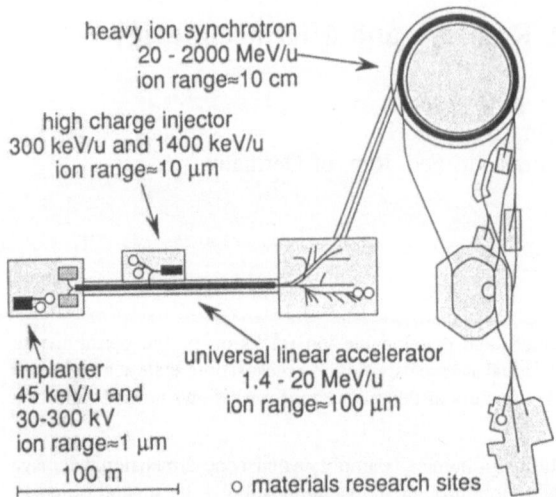

heavy ion synchrotron
20 - 2000 MeV/u
ion range≈10 cm

high charge injector
300 keV/u and 1400 keV/u
ion range≈10 μm

implanter
45 keV/u and
30-300 kV
ion range≈1 μm

universal linear accelerator
1.4 - 20 MeV/u
ion range≈100 μm

100 m

o materials research sites

Figure 1. Heavy ion accelerator facilities, materials research sites, and corresponding ion ranges in solids at GSI.

realm. Sixty years later, in 1958, Young observed the first persistent tracks in solids [2]. While tracks in fluids are transient phenomena of short duration, wiped-out after fractions of a second, tracks in solids may endure millions of years, often without perceivable fading. Thus, the „frozen" damage of fossil tracks in solids is capable to reveal events from the far back history of our world [3]. Significant scientific topics are to study the thermal history and calibrate time scales in geology and astrophysics, and to study exotic fission and shock waves in nuclear and high energy physics. Gradually the diagnostic concept of track observation was complemented by that of a structural tool for imprinting structure onto solids [4], [5].

The report presented here is strongly based on activities at the heavy ion accelerator facility GSI [6], [7] (Figure 1) in collaboration with other laboratories. The irradiation facilities at GSI comprise experimental sites with ion ranges in solids up to more then 10 centimeters. Most sites enable a wide-beam operation with several centimeters beam diameter. In addition, an ion microbeam facility with a 1 μm focal diameter exists [8]. In contrast to many basic science experiments, materials research irradiations require a precise control of the beam homogeneity and of the accumulated fluence, i.e. the number of ion tracks per unit area. While at low ion energies stoichiometric applications changing the structure and composition of solids prevail, at high energies structural modifications — such as ion tracks — are more frequently induced.

Basic features of ion tracks

Creation of defects

On the way from the ion irradiation to the observation of developed tracks three distinct levels of complexity exist, associated with different time scales (Figure 2). The goal of basic track research is to understand the spatial distribution and density of the created effect, the track etch threshold and the saturation of the effect with increasing energy input along the ion path.

During its passage through the solid the ion imparts its energy gradually in small portions to the quasi-free electrons of the solid, leading to a long range electronic collision cascade corresponding to the track halo. In insulating solids the resulting positive space charge distribution leads to a short

2

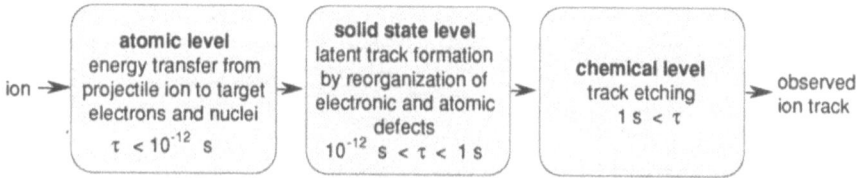

Figure 2. Basic steps in ion track creation.

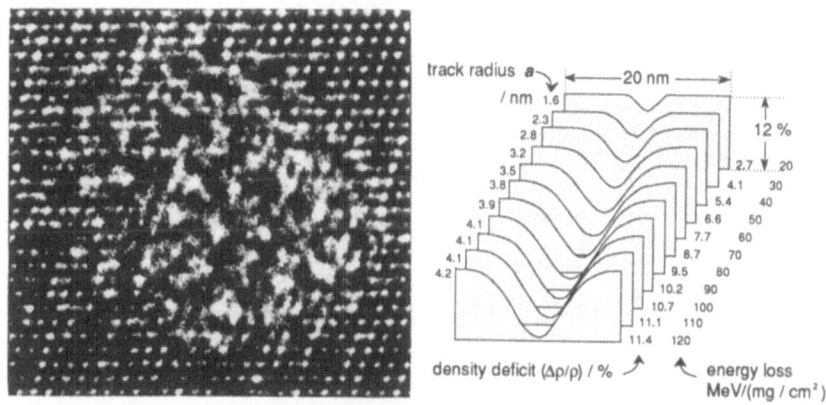

Figure 3. Latent track structure observed by transmission electron microscopy (left) [12] and by neutron small angle scattering (right) [13], [14], [15], [16], [17].

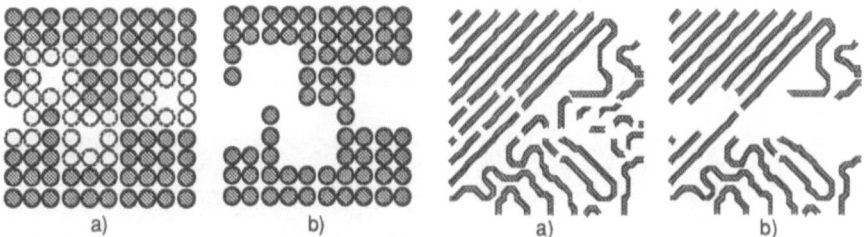

Figure 4. Preferential dissolution of defect zones in crystal (left) and polymer (right).

range atomic collision cascade corresponding to the track core. Besides in insulating solids tracks have been even found in metallic crystals [9], [10], [11]. Along the ion path stable defect agglomerates (extended defects) are created, chemically active, amorphous zones of roughly 10 nm diameter with a decreased density similar to that of a glassy phase (Figure 3).

The linear density of the extended defects defines their mutual overlap and thus the etchability of the resulting latent track. Latent tracks are zones of increased mobility susceptible to a preferential dissolution in the etch bath (Figure 4).

Enhancement and annealing of defects

In the vicinity of the track etch threshold already small changes in the diameter of the extended defects have a large influence on the track etch speed. Various techniques can be used to increase or decrease the size of the extended defects before their ultimate chemical etching. In the case of polymers uv treatment or soaking in a suitable organic solvent [18], [19] before etching can improve track etchability. On the other hand, annealing is usually achieved by a thermally activated diffusion process.

Selective chemical etching of latent ion tracks

The shape of the etched track depends mainly on the ratio of the track etch-rate in the activated zone and of the bulk etch-rate in the undamaged solid. Diffusion and convection play the dominant role for short and long-range transport of the reaction products, respectively (Figure 5 left). In polymers the reaction zone consists of a gradually depolymerizing layer forming a continuous transition of chain sizes between the virgin polymer and the solution (Figure 5 right). On the molecular level the track etch rate is determined by the number of reactive sites in the form of chain ends, by the so-called „free volume" in the vicinity of the reactive sites accessible to the etchant during its approach to the reactive site, by the reaction rate of the chemical reaction, and by the required degradation of the polymer before it is ultimately dissolved. Typical track shapes are shown in Figure 6.

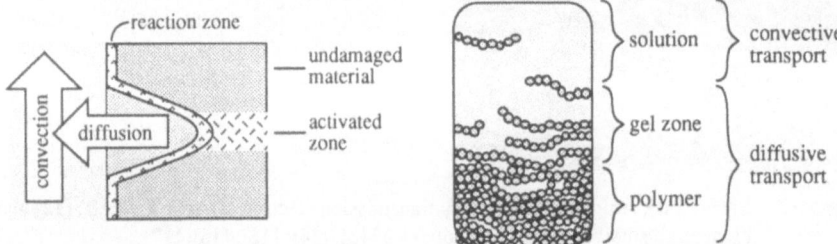

Figure 5. Primary factors in track etching. The etching of a latent ion track depends on the density of the radiation damage along the ion track, on the reaction rate between the solid and the etchant, and on the rate of removal of the etch products by diffusion and convection (left). Reaction zone in polymer etching (right).

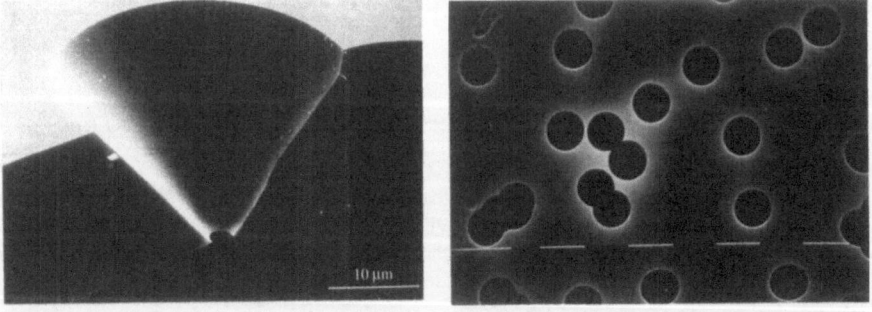

Figure 6. Conical track in boron glass corresponding to small track etch ratio (left) and cylindrical tracks in polycarbonate corresponding to high track etch ratio (right).

Suitable parameters for influencing the results

The track technique is ruled by the range of the ion, the angle of incidence, the diameter and density deficit of the latent track with respect to the bulk material (Figure 7 left), and the areal density of tracks (Figure 7 right). The possibility of a precise control of these parameters in connection with

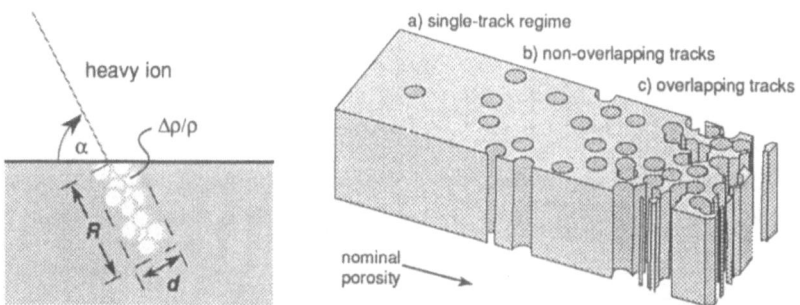

Figure 7. Basic parameters of tracks (left) and track density regimes (right).

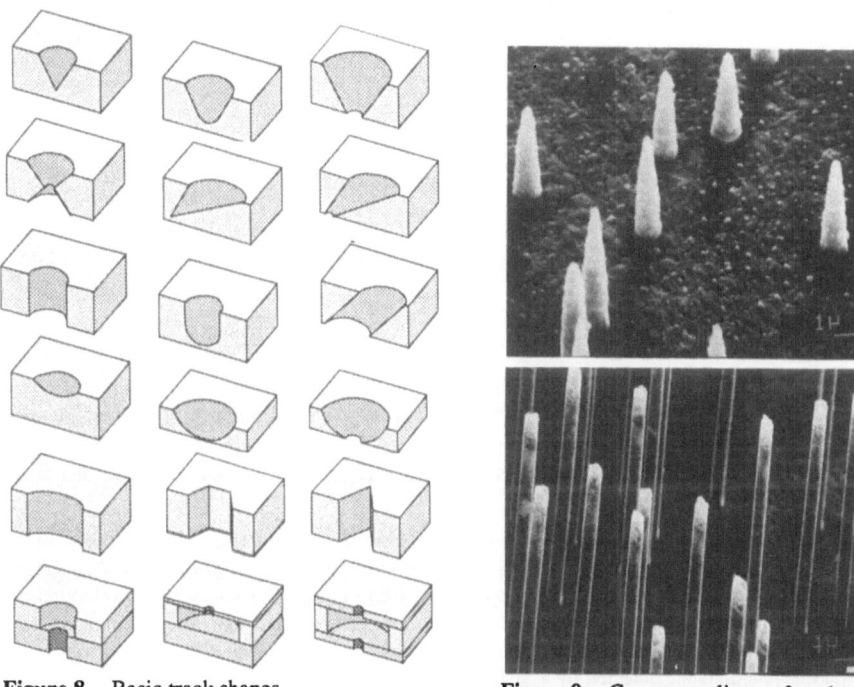

Figure 8. Basic track shapes.

Figure 9. Copper replicas of etched tracks in polycarbonate (top) and in muscovite mica (bottom).

track etching is a key feature for practical applications. Figure 8 gives an overview on basic shapes obtained by variation of the available parameters. Replica techniques can be used to transfer these shapes to other materials (Figure 9).

Ion track applications

Single pores in medicine and low temperature physics

The characterization of suspended particles in fluids is of primary interest in several fields, ranging from water pollution to medical diagnostics. Among the most vitally required data are the

5

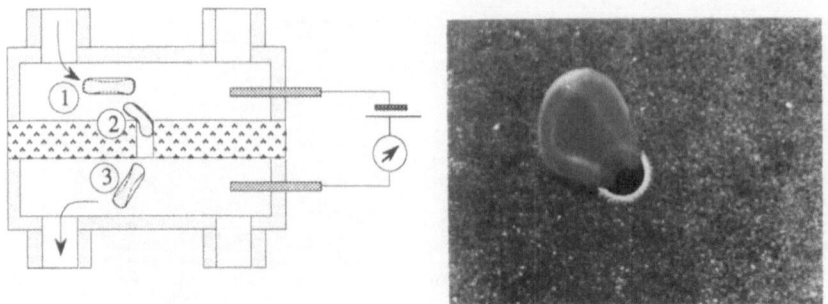

Figure 10. Red blood cell counter for characterizing the deformability of individual cells (left) and cell passage (right).

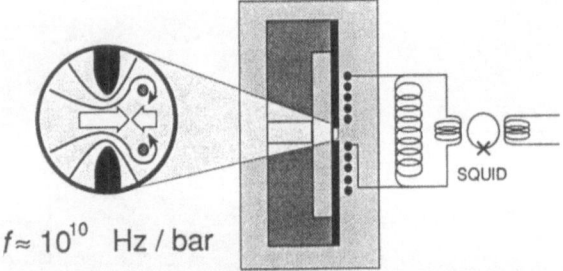

$f \approx 10^{10}$ Hz / bar

Figure 11. Principle of superfluid pressure sensor translating pressure differences into a frequency.

number-density of the particles, their volume, and their deformability. Single-ion tracks [20] enable to define a deformability parameter for individual red blood cells (Figure 10) providing a new technique for medical diagnostics and the development of drugs related to membrane deformability [21], [22], [23].

Ion tracks provide obstacles sufficiently small to interfere with superfluids [24]. The Josephson effect in suprafluid helium [25] is based on the weak coupling of two adjacent suprafluid reservoirs. It relates a pressure difference with a frequency and promises the development of highly sensitive sensors for pressure [26] and rotation. Above a critical flow a quantum vortex is created (Figure 11 left) corresponding to an elementary frictional event that shifts the membrane by a discrete distance detected by a SQUID sensor (Figure 11 right).

Multiple pores in filtration and membrane technology

Almost immediately after the first observation of etched ion tracks in mica, the potential of ion track filters for the mechanical separation of small particles suspended in a fluid was recognized [27]. Ion track filters are defined by very few, almost independent parameters, the length, diameter, and areal density of the tracks. These parameters can be varied over several orders of magnitude and result in stochastic track patterns with predictable properties [28], [29], [30]. Possibilities comprise asymmetric pores for cross-flow filtration [31], pore wall modifications, and isotope separation [32] (Figure 12). Very small etched tracks and latent tracks show enhanced diffusion [33], [34] and could be used for the separation of gases in the future.

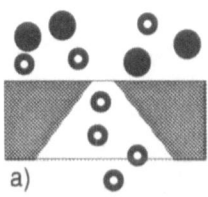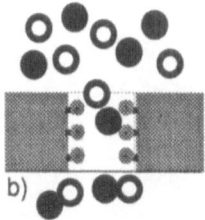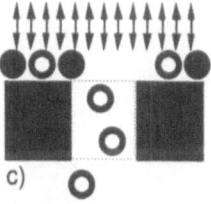

Figure 12. Membrane technology. (a) Asymmetric pores for cross-flow filtration. (b) The possibility of modified pore walls for encymatic reactors. (c) Isotope separation by laser-induced surface desorption.

Creation of directed (anisotropic) properties in solids

The influence of ion tracks on the magneto-optic properties of iron garnets represents a study case how bulk properties can be influenced and directionally dependent properties created by ion tracks [35], [36], [37]. Ion tracks reduce the wall energy of magnetic domains along the direction of the irradiation and lead to a directional pinning of the domain walls (Figure 13).

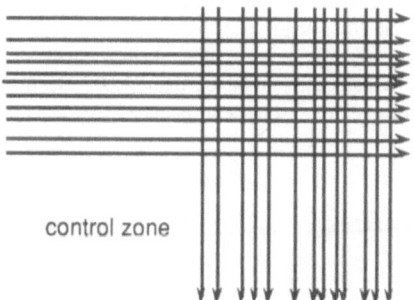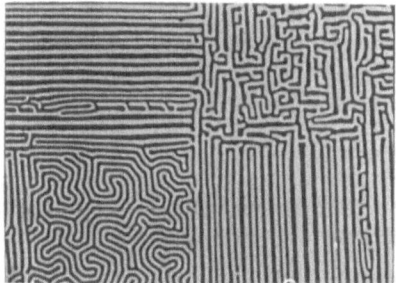

Figure 13. Influence of oblique ion tracks on magnetic domains. The direction and area of the irradiation is indicated by arrows (left). Clockwise, four zones can be distinguished (right): A zone with horizontal domain pattern, a zone with both, horizontal and vertical domain pattern, a zone with vertical domain pattern, and finally a control zone with isotropic domain pattern.

Prospects of heavy-ion lithography

At high track densities, ion tracks can be utilized for revealing projected areal densities of objects (Figure 14 left) and for imprinting structures onto solids (Figure 14 right) [38], [39], [40]. The technique has several new features in comparison with conventional techniques. It is a single particle tool in which each ion creates exactly one latent particle track and leads to exactly one characteristic hollow shape during the etching. Besides light-sensitive photoresists the technique enables to shape a wide range of materials. Due to the small angular straggling of heavy ions in a light matrix very deep structures can be obtained. The finest structures obtained until now are channels of approximately 0.01 μm diameter. The depth of the resulting structure can be controlled by the energy of the incident ions. Directionally dependent properties can be created in solids. The accessible radiation sources provide intense ion beams with high parallelism.

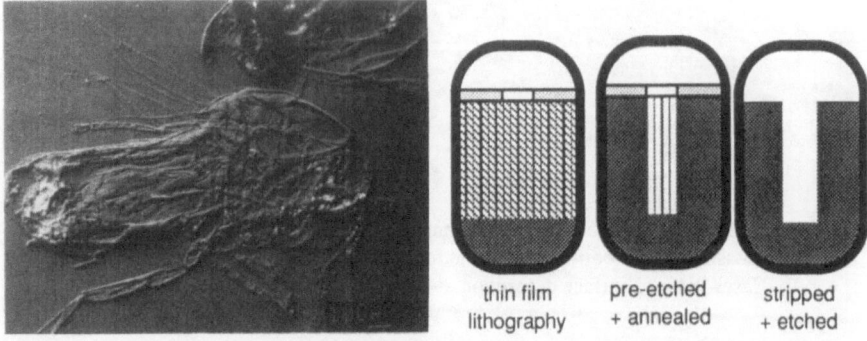

thin film   pre-etched   stripped
lithography   + annealed   + etched

Figure 14. Heavy ion lithography. (left) Ion lithogram of insect. The relief elevation is proportional to the projected areal density of the object. (right) Scheme for generating deep structures by conventional lithography in connection with an ion irradiated substrate and an annealing step.

Low energy ion tracks for generating surface texture

A little noticed aspect of the ion track technique is the possibility to generate directed surface textures with precisely defined dimensions. One possibility is to generate light scattering devices using a large number of conical tracks spreading an incident light beam into a well-defined solid angle [41]. At high density of the ion irradiation such a surface will have a decreased reflection (Figure 15) [42]. Another possibility is to increase the electrical surface resistance by increasing the surface roughness [43].

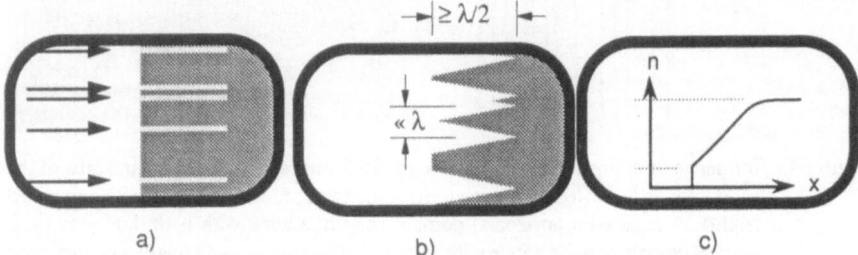

a)   b)   c)

Figure 15. Principle of antireflection treatment of surfaces. The sample is irradiated at a high ion fluence (a), so that neighboring tracks will be at a distance closer than the wavelength of the used radiation (b). The resulting density and refractive index (c) therefore represents a smooth transition from the surface to the bulk material. Thereby the reflection can be reduced — depending on the wave length — by orders of magnitude.

Focussed ion beams for creating predetermined shapes

Ion microbeams provide the possibility to generate predefined track patterns in solids. Such a microtool [44] provides the versatility of a computer-controlled mechanical lathe with a lateral resolution of about 1 μm and a penetration depth of about 100 μm (Figure 16) [45].

Perspective

The sweeping progress of ion source and accelerator technology is a challenge for scientists and engineers to promote as well their practical uses. While low-energy ion implantation is already established as a standard tool in semiconductor and surface technology, at considerably higher ion

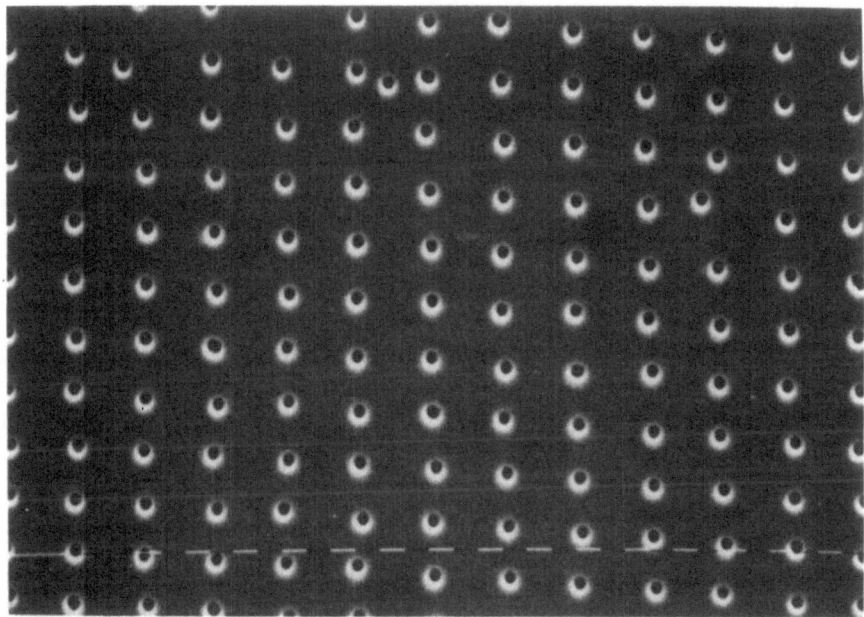

Figure 16. Regular pattern of etched single-ion tracks [46]. The conical tracks are inscribed with ^{52}Cr ions of 1.4 MeV/u specific energy. Normal soda-lime glass was used as a track recorder. The etched track cones have an opening angle of about 70°. The lattice has a characteristic distance of ca. 30 μm. Less than 1 % scattered particles are observed. Bar length = 10 μm.

energies the phenomenon of discrete and directed ion tracks appears. This opens the possibility to structure and modify solids with individual atomic particles.

Characteristic properties of ion tracks

- **Single-particle tool**

  Each ion creates exactly one latent particle track and leads to exactly one characteristic hollow shape during the etching.

- **Wide range of eligible materials**

  Besides light-sensitive photoresists, many radiation resistant polymers, glasses, and crystals can be used.

- **High depth of resulting structures**

  Due to the small angular straggling of heavy ions in a light matrix very deep structures can be obtained.

- **Controlled depth of structuring**

  The depth of the resulting structure is determined by the energy of the incident ions. Range straggling is small.

- **High lateral resolution**

  The finest structures obtained until now are channels of approximately 0.01 μm diameter.

- **Generation of directed properties (anisotropy) possible**

  Directionally dependent properties can be created in solids by the collective action of many ion tracks.

- **High quality of available radiation sources**

  Ion accelerators are able to provide intense ion beams with high parallelism.

9

References

[1]   Wilson, C.T.R., Phil.Trans. A189, 265, (1897).

[2]   Young, D.A.: "Etching of Radiation Damage in Lithium Fluoride." Nature 182, 375-377 (1958).

[3]   Fleischer, R.L., P.B. Price, R.M. Walker: "Nuclear Tracks in Solids: Principles and Applications." University of California, Berkeley, 1-605, (1975).

[4]   Fischer, B.E., R. Spohr: "Production and Use of Nuclear Tracks: Imprinting Structure on Solids." Rev.Mod.Phys., 55, No.4, 907-948, (1983).

[5]   Spohr, R.: "Ion Tracks and Microtechnology. Basic Principles and Applications." Vieweg, 272 pp. (1990).

[6]   Angert, N.: "Ion Beams at GSI." 15th Int. Conf. on Particle Tracks in Solids, Marburg, Sept. 3-7, (1990).

[7]   Rück, D.M., N. Angert, H. Emig, K.D. Leible, P. Spädtke, D. Vogt, B.H. Wolf: "Ion Implantation Facilities at GSI." 15th Int. Conf. on Particle Tracks in Solids, Marburg, Sept. 3-7, (1990).

[8]   Fischer, B.E.: "The Heavy Ion Microprobe at GSI — Used for Single-Ion Micromechanics." Nucl. Instr. Meth. in Physics Research, B 30, pp. 284-288, (1988).

[9]   Dunlop, A., D. Lesueur, J. Morillo, J. Dural, R. Spohr, J. Vetter: "High Electronic Excitations and Damage Production in GeV Ion Irradiated Metals." Nuclear Instruments and Methods in Physics Research, B48, pp. 419-424, (1990).

[10] Adouard, A., E. Balanzat, S. Bouffard, J.C. Jousset, A. Chamberod, A. Dunlop, D. Lesueur, G. Fuchs, R. Spohr, J. Vetter, L. Thomé: "Evidence for Amorphization of a Metallic Alloy by Ion Electronic Energy Loss." Physical Review Letters, vol 65, No. 7, pp 875 - 878, (1990).

[11] Barbu, A., A. Dunlop, D. Lesueur, R.S. Averback, R. Spohr, J. Vetter: "First Transmission Electron Microscopy Observation of Latent Tracks in a Metallic Compound." Europhysics Letters vol. 15, 37, (1991).

[12] Narayan, J, Oak Ridge National Laboratory, P.O. Box X, Oak Ridge, Tennessee, 37830/ USA.

[13] Albrecht, D., P. Armbruster, M. Roth, R. Spohr: "Small Angle Neutron Scattering Observations from Oriented Latent Nuclear Tracks." Supplement No. 3 to Nuclear Tracks, pp. 55 - 58, (1982).

[14] Spohr, R., P. Armbruster, K. Schaupert: "Structure and Diffusion Properties of Latent Ion Tracks." Radiation Effects and Defects in Solids, 110, Nos. 1-2, pp. 27-31, (1989).

[15] Albrecht, D., P. Armbruster, M. Roth, R. Spohr: "Small Angle Neutron Scattering Observations from Oriented Latent Nuclear Tracks." Radiation Effects, vol. 65, pp. 145 - 148, (1982).

[16] Albrecht, D., P. Armbruster, R. Spohr, M. Roth, K. Schaupert, H. Stuhrmann: "Small Angle Scattering from Oriented Latent Nuclear Tracks." Nucl. Instrum. and Meth., B2, vol. 230, no.1-3, pp. 702 - 705, (1984).

[17] Albrecht, D., P. Armbruster, R. Spohr, M. Roth, K. Schaupert, H. Stuhrmann: "Investigation of Heavy Ion Produced Defect Structures in Insulators by Small Angle Scattering." Appl. Phys., A 37, pp. 37 - 46, (1985).

[18] Lück, H.B., H. Matthes, B. Gemende, B. Heinrich, W. Pfestorf, W. Seidel, S. Turuc: "Production of Particle-Track Membranes by Means of a 5 MV Tandem Accelerator." Nucl. Instr. Meth. in Phys. Res. B50, pp. 395 - 400, (1990).

[19] Lück, H.B.: "Solvent-Induced Sensitization of Particle Tracks in Polyester." 15th Int. Conf. on Particle Tracks in Solids, Marburg, Sept. 3-7, (1990).

[20] Roggenkamp, H.G., H. Kiesewetter, R. Spohr, U. Dauer, L.C. Busch: "Production of Single Pore Membranes for the Measurement of Red Blood Cell Deformability." Biomedizinische Technik, vol. 26, pp. 167 - 169 (1981).

[21] Kiesewetter, H., K. Mussler, P. Teitel, U. Dauer, H. Schmid-Schönbein, R. Spohr: "New Methods for Red Cell Deformability Measuremen." Clinical Aspects of Blood Viscosity and Cell Deformability, Springer, Berlin, pp. 19 - 26, (1981).

[22] Guillet, R., J. Vetter, D. Koutsouris, Y. Beuzard, M. Boynard: "Individual Red Blood Cell (RBC) Transit Analysis Through Sized Micropores of Oligopore Filters: Application to RBC Subpopulation. " 15th Int. Conf. on Particle Tracks in Solids, Marburg, Sept. 3-7, (1990).

[23] Seiffge, D.: "Application of Single-Pore Membranes in Hemorheological Research." 15th Int. Conf. on Particle Tracks in Solids, Marburg, Sept. 3-7, (1990).

[24] Pekola, J.P., J.C. Davis, Zhu Yu-Qun, R.N.R. Spohr, P.B. Price, R.E. Packard: "Suppression of the Critical Current and the Superfluid Transition Temperature of ^{3}He in a Single Submicron Cylindrical Channel." Journal of Low Temperature Physics, 67, 47 (1987).

[25] Varoquaux, E., O. Avenel: "Quantum Phase Slippage in Superfluid 4He." Physica Scripta, vol. T19, pp. 445 - 452, (1987).

[26] Eska, G., Y. Hirayoschi, C. Trautmann, R. Spohr, J. Vetter: "Pressure and Rotation Sensors Based on the Superfluid Josephson Effect." GSI Report 91-1, p. 260, (1991).

[27] Price, P.B., R.M. Walker: "Molecular Sieves and Methods for Producing Same." United States Patent Office, No. 3,303,085, Feb. 7, (1967)

[28] Riedel, C., R. Spohr: Statistical Properties of Etched Nuclear Tracks. I. Analytical Theory and Computer Simulation." Radiation Effects, vol. 42, pp. 69 - 75, (1979).

[29] Riedel, C., R. Spohr: "Statistical Properties of Etched Nuclear Tracks II. Experiment and Filter Design." Radiation Effects, vol. 46, pp. 23 - 30, (1980).

[30] Riedel, C., R. Spohr: "Transmission Properties of Nuclear Track Filters." J. Membrane Science, vol. 7, pp. 225 - 234, (1980).

[31] Lück, H.B., B. Gemende, B. Heinrich: "Structure Modification of Particle Track Membranes." 15th Int. Conf. on Particle Tracks in Solids, Marburg, Sept. 3-7, (1990).

[32] Kravchenko, V.A.: "Laser Induced Desorption from Metallized Microporous Membranes." 15th Int. Conf. on Particle Tracks in Solids, Marburg, Sept. 3-7, (1990).

[33] Schaupert, K., D. Albrecht, P. Armbruster, R. Spohr: "Permeation Through Latent Nuclear Tracks in Polymer Foils." Appl. Phys. A 44, 347-352 (1987).

[34] Spohr, R., P. Armbruster, K. Schaupert: "Structure and Diffusion Properties of Latent Ion Tracks." Radiation Effects and Defects in Solids, vol. 110, Nos. 1-2, pp. 27-31, (1989).

[35] Krumme, J.P. I. Bartels, B. Strocka, K. Witter, Ch. Schmelzer, R. Spohr: "Pinning of 180° Bloch Walls at Etched Nuclear Tracks in LPE-Grown Iron Garnet Films." Applied Physics, 48, 5191-5196 (1977).

[36] Hansen, P. H. Heitmann: "Influence of Nuclear Tracks on the Magnetic Properties of a (Gd,Bi)3(Fe,Ga)5O12 Garnet Film." Phys.Rev.Lett. 43, 1444-1447 (1979).

[37] Hansen, P., H. Heitmann, B. Strocka, R. Spohr: "Der Einfluß von Ionenstrahlen auf die magnetischen Eigenschaften von ferrimagnetischen Schichten." BMFT Bericht FB T83-048 Fachinformationszentrum Karlsruhe, D-7514 Eggenstein-Leopoldshafen 2, 1-180 (1983).

[38] Yang, T.C., G. Welch, C.A. Tobias, H. Maccabee, T. Hayes, L. Craise, E.V. Benton, F. Abrams, Annals of the New York Academy of Sciences, 306, pp. 3322-339, (1978)

[39] Fischer, B.E., B. Genswürger, R. Spohr: Heavy Ion Lithography: "A Tool for Investigation and Replication of Microscopic Objects." Int. J. Appl. Rad. Isotopes, vol. 31, pp. 297 - 305, (1980).

[40] Fischer, B.E., R. Spohr: "Heavy Ion Microlithography — a New Tool to Generate and Investigate Submicroscopic Structures." Nuclear Instruments and Methods, 168, 241-246, (1980).

[41] Buerckner, D., E. Pfeng, T. Popp, R. Spohr: "Optical Refraction by Etched Nuclear Tracks." GSI report 86-1, p. 268, (1986)

[42] Fischer, B.E.: "Graded-Index Antireflecting Surfaces." GSI 84-1, ISSN 0174-0814. p. 216, (1984).

[42] Fischer, B.E., R. Spohr: "Preparation of Superinsulating Surfaces by the Nuclear Track Technique." Radiation Effects, vol.65, pp. 143 - 144, (1982).

[43] Fischer, B.E., Gesellschaft für Schwerionenforschung mbH, Postfach 110 552, D-6100 Darmstadt, FRG

[45] Fischer, B.E.: "The Scanning Heavy Ion Microscope at GSI." Nucl. Instr. Meth. in Physics
      Research, B 10/11, 693, (1985).

[46] Fischer, B.E.: "The Heavy Ion Microprobe at GSI — Used for Single-Ion Micromechanics."
      Nucl. Instr. Meth. in Physics Research, B 30, pp. 284-288, (1988).

Rare Gases in Metals – Influence on the Formation and Nucleation of Cavities

R. Schumacher and R. Vianden

Institut für Strahlen- und Kernphysik, Universität Bonn,
Nußallee 14–16, W-5300 Bonn, Fed. Rep. of Germany

The nucleation and growth of cavities at In impurities in metals in the presence of implanted rare gases is studied by means of the γ-γ perturbed angular correlation method. From the properties of the measured electric field gradients it is concluded that the In probe atoms are incorporated in the inner surfaces of the cavities. A survey of the results obtained in the fcc metals Al, Ni, Cu and Au as well as the bcc metals Mo and W is given. A linear variation of the elctric field gradient with the surface temperature was found in agreement with theoretical predictions. Further it could be shown that the heavy rare gases Kr and Xe diffuse in Cu via the usual vacancy type mechanism whereas for He and Ne the so called "dissociative mechanism" must be operative.

1. Introduction

The behaviour of rare gases in metals was long time only of theoretical interest since classical metallurgical methods do not allow the formation of any alloy between a metal and a rare gas. For the solubility of He in Ni e.g. an upper limit of 10^{-7} at. % at 1373 K and 100 bar He pressure could be established experimentally [1].

With the advent of nuclear technology and quite recently the application of thin films produced by ion beam mixing or sputtering with Xe or Ar beams this attitude changed. In fission reactors n,α reactions lead to the formation of isolated He atoms in the metallic lattice of the structural components and their interaction with other defects in the lattice causes swelling, embrittlement and ultimately the failure of the components. In planned fusion reactors the rate of He introduction in the structural components is about 10 - 100 times higher and therefore these problems play an even more important role.

In order to understand and possibly prevent the detrimental effects of He in metals it is necessary to understand the details of the interaction of rare gases with the undisturbed metal lattice and other intrinsic or impurity atom defects. The electric field gradient (EFG) at the site of a lattice atom is extremely sensitive to deviations from the perfect cubic lattice symmetry in the nearest neighbourhood. This is the reason why the **p**erturbed γ - γ **a**ngular **c**orrelation method (PAC) which allows the determination of the EFG at the site of a suitable probe nucleus is a perfect method to study these questions. In the past the PAC method has significantly contributed to our understanding of the nature and behavior of defects in metals and semiconductors [2, 3].

Recently PAC investigations in Cu led to the conclusion that in the presence of rare gases the usual two-dimensional agglomerations of vacancies (vacancy loops) are changed to threedimensional cavities. The PAC probe ^{111}In allowed the study of the first stages of the nucleation and growth of this cavities [4]. Here after a short description of the principles of the PAC method the results obtained for the formation process and the behaviour of In in the inner surfaces of cavities in Cu as well as Al, Au, Ni, Mo and W will be described.

2. Experimental details

The perturbed $\gamma-\gamma$ angular correlation method is based on the fact that the probability of photon emission from an exited nuclear state depends on the angle between the emission direction and the nuclear spin. If a nucleus decays by the successive emission of two γ - rays one can exploit this by measuring the emission probability of the second γ ray (γ_2) relative to the direction in which the first (γ_1) was detected. The presence of electromagnetic fields at the site of the nucleus during the lifetime of the intermediate state of the $\gamma-\gamma$ cascade leads to periodic changes in the emission probabilty of γ_2 in a fixed direction relative to the emission direction of γ_1. Semiclassically this process can be pictured as caused by the precession of the nuclear spin about the axis of the electromagnetic field. In defect studies one usually exploits the electric quadrupole interaction between the nuclear quadrupole moment and an electric field gradient (EFG) at the site of the nucleus. A charge distribution of cubic or higher symmetry does not produce an EFG at the origin. If however this symmetry is broken, e.g. by the presence of a defect in the nearest neighbourhood of a substitutional probe atom in a cubic lattice, this defect gives rise to an EFG at the site of the nucleus of a PAC probe atom which can be detected by the effects on the $\gamma-\gamma$ angular correlation. From the measurements one can derive the quadrupole interaction frequency:

$$\nu_Q = e \cdot Q \cdot V_{zz} / h$$

with the nuclear quadrupole moment Q and the main component of the EFG tensor V_{zz}. The shape of the perturbation function allows further the determination of the symmetry of the EFG tensor, given by the asymmetry parameter η. For an easier comparability of the results in the discussion and the tables only the quadrupole precession frequency ω is given which is related to ν_Q in the following way:

$$\omega = \pi/10 \cdot \sqrt{3} \cdot \alpha \cdot \nu_Q \cdot \sin(1/3 \cdot \arccos \beta)$$

$$\alpha = \sqrt{28/3 \cdot (3+\eta^2)} , \; \beta = 80 \cdot (1 - \eta^2) / \alpha^3$$

Measurements in single crystals yield the orientation of V_{zz}, the largest component of the EFG tensor, relative to the crystal axes of the host lattice.

The PAC probe atom used in the following investigations was ^{111}In which decays via a 173 - 247 keV $\gamma-\gamma$ cascade to the ground state of [111]Cd.

In the case of Mo and W single crystals were used for the measurement of the EFG orientation. The ^{111}In probe atoms were implanted with energies between

Table 1: Properties and implantation parameters of metal samples

	sample purity and thickness	^{111}In-implantation		postimplantation		
		energy [keV]	$R_p \pm \Delta R_p$	element, energy [keV]	$R_p \pm \Delta R_p$ [Å]	dose [I/cm²]
Cu	99.999 at.% 25 μm	80	164 ± 71	He, 11	585 ± 302	$2 \cdot 10^{14}$
				Ne, 18	169 ± 95	
				Ar, 25	129 ± 69	
				Kr, 80	198 ± 94	
				Xe, 80	149 ± 59	
				N, 13	173 ± 94	
Ni	99.99+ at.% 25 μm	160	272 ± 114	He, 10	479 ± 240	$5 \cdot 10^{14}$
				Ne, 33	273 ± 141	$2 \cdot 10^{14}$
				Ar, 60	265 ± 127	
				Kr, 125	276 ± 122	
				Xe, 160	240 ± 89	
Al	99.99 at.% 12.5 μm	80	411 ± 124	Ar, 35	379 ± 148	$2 \cdot 10^{14}$
Au	99.9 at.% 25 μm	160	230 ± 133	Ar, 50	212 ± 122	
Mo	99.9 at.% 25 μm	80	169 ± 84	Ar, 30	161 ± 90	
W	99.95 at.% 10 μm	80	128 ± 74	Ar, 35	150 ± 87	
Cu	99.999 at.% 25 μm	140	257 ± 110	Ne, 28	251 ± 133	$3 \cdot 10^{14}$
				Kr, 105	253 ± 117	$2.5 \cdot 10^{14}$
				Xe, 160	253 ± 98	$2 \cdot 10^{14}$

80 and 160 keV and doses of typically 10^{13} at/cm². Subsequently the desired rare gas species was implanted with doses of ~ 2 x 10^{14} at/cm² and energies adjusted to achieve a maximum overlap with the In profile. In the experiments aimed at the study of the growth mechanism for cavities in Cu the doses of the rare gas implantations were varied for Ne, Kr and Xe between 2 and 3 x 10^{14} at/cm² in order to get identical rare gas concentrations in the peak of the Indium range profile. All implantations were carried out at room temperature. The properties and implantation parameters for typical samples are collected in table 1.

After the rare gas implantations the variation of the PAC signal was studied in an isochronous annealing program. The annealing steps were carried out in a vacuum of p ≤ 10^{-6} kPa in and above the temperature range of annealing stage III of the different materials. The holding time was 10 min. Base temperature of the program was 293 K except in the case of Ni where due to the ferromagnetism the measurements were carried out at a sample temperature of 690 K, i.e. above the Curie temperature.

3. Results and discussion

a. Origin of the EFGs observed in Cu, Ni, Al, Au, Mo, W

In Cu and Ni the development of the PAC signal was studied after implantation of all stable rare gases. It was found in both metals that the presence of the rare gases led to the appearance of a new characteristic quadrupole precession frequency ω that is not found in samples implanted with metal ions at the same dose (Fig. 1). The magnitude of the frequency is independent of the implanted gas species.

In Cu the value of ω at 293 K increased by a few percent from the lowest to the highest annealing temperatures. This increase was accompanied by a decrease of η from $\sim$ 0.2 to values below 0.1 and a reduction of the damping of the PAC pattern. In Cu identical results were found after the implantation of N (see Tab. 1).

In Ni the characteristic quadrupole precession frequency found after rare gas implantations has values of $\omega \sim$ 220 - 230 Mrad/sec measured at sample temperatures between 700 and 850 K. The value of η is 0.15 [5]. This interaction has already been observed by Pleiter et al. [6] after the postimplantation of He and the orientation of the V_{zz} component of the EFG was determined to be parallel to the $\langle 111 \rangle$ crystal direction.

Due to the fact that all rare gases led to the same results in the fcc metals Cu and Ni, for Al and Au only the effects of an Argon postimplantation were studied. Again the dominating frequencies which appeared were already known from He postimplantation experiments [7, 8]. For Au the value is ω = 212 Mrad/sec and η = 0 and for Al ω = 138 Mrad/sec, η = 0. In Al ω was found to decrease untypically to 128 Mrad/sec after annealing at 773 K. Both V_{zz} components showed $\langle 100 \rangle$ orientations.

The results for the bcc metals Mo and W differed in one fundamental point from those obtained for the fcc metals. Here the rare gas postimplantations were not a necessary condition for the appearance of characteristic frequencies at the end of annealing stage III. The parameters are ω = 212 Mrad/sec, η = 0.15 and ω = 248 Mrad/sec, η = 0 for Mo and W, respectively. The orientation of V_{zz} is parallel to $\langle 100 \rangle$ in both metals. It is interesting to note that the frequencies disappeared if the samples in which they were already present were postimplanted with Ar, but reappeared after annealing at the same temperatures than before.

As an explanation for the various effects observed in Cu and some of the metals mentiond above after He implantation the formation of "small He filled vacancy clusters consisting of 2 - 3 vacancies and 2 - 3 He atoms" was proposed [9, 6]. However this model fails to explain:

- the newly found independence of ω on the rare gas species, since the different sizes of e.g. He and Xe should lead to different relaxations in the neighbourhood of the probe atoms and thus to considerably different EFGs due to its $1/r^3$ dependence.

- the gradual changes of ω and η observed during the annealing programs are difficult to understand since a relatively small complex has only a few well defined

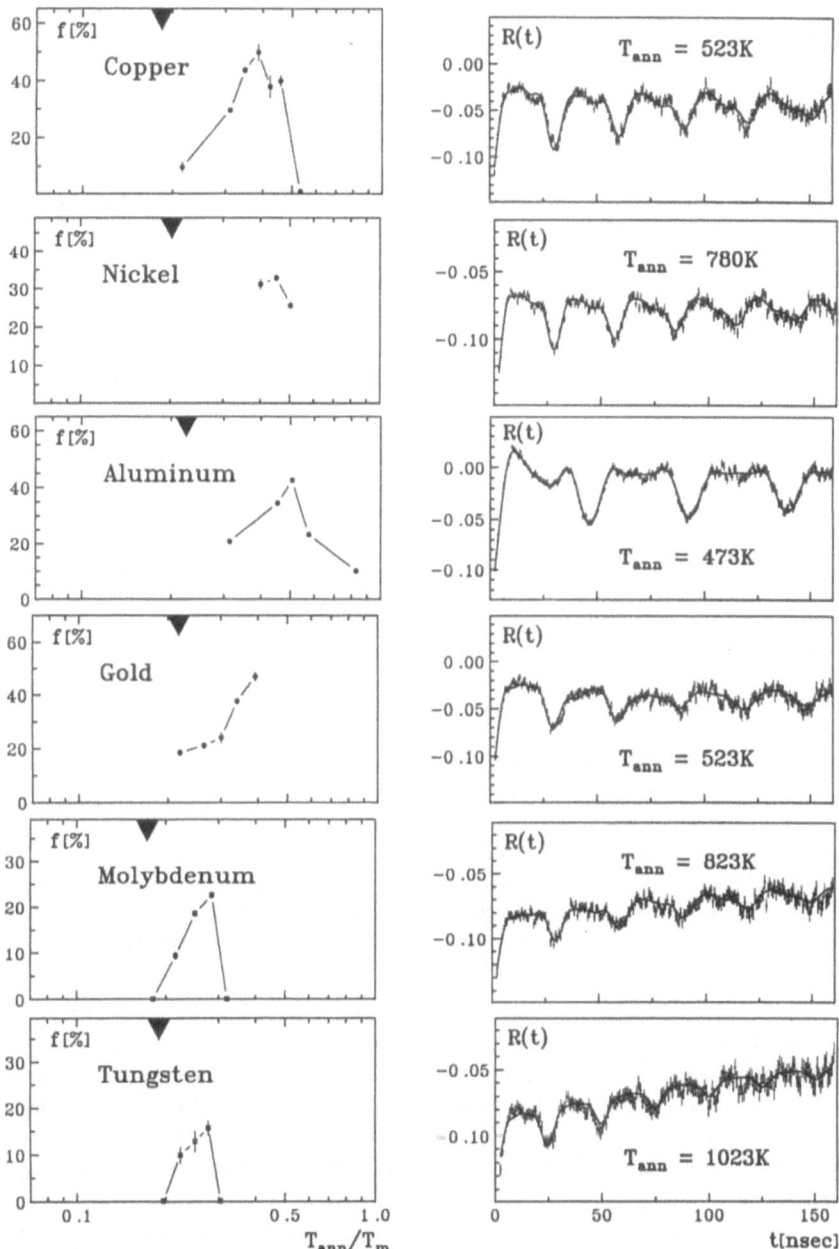

Figure 1: Fraction of ^{111}In probe atoms in the inner surface of cavities in different metals after annealing (left). The annealing temperature T_{ann} is normalized to the melting temperature T_m. The filled triangle at the upper frame of the plot indicates the position of annealing stage III. Typical PAC spectra corresponding to each host metal are shown on the right.

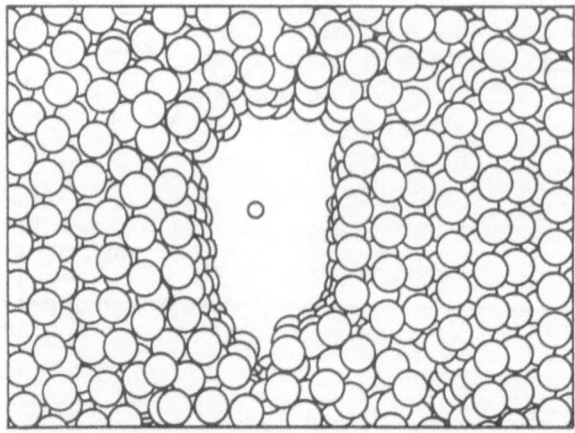

Figure 2: Artists view of a cavity containing one rare gas atom. The ^{111}In probe atom would be located in the inner surface bounding the cavity.

possible configurations. In one case, Al, the configuration suggested by Pleiter et al. [7] could even be disproved by lattice location channeling experiments [10].

- such different configurations would lead to several well defined precession frequencies. Therefore the appearance of just one ω is quite unexpected.

These difficulties could be resolved by a new model suggested by Schumacher and Vianden [4]. There it is assumed that the observed quadrupole interaction frequencies are due to the electric field gradient experienced by probe atoms in the inner surfaces of large rare gas filled vacancy clusters (cavities) as shown in figure 2. A few rare gas atoms then only catalytically contribute to the growth of the cavities and prevent them from collapsing to two-dimensional vacancy agglomerates (loops) which are well known [11, 12] in the fcc metals. In this picture all the inconsistencies mentioned above disappear.

Especially the well defined EFG and the irreversible shifts of the EFG during the annealing program can easily be understood to be due to the growth of the cavities by collecting more vacancies set free from shallow traps. This process moves the irregularities of the cavities further away and thus reduces their disturbing influence on the surface EFG. The neutral rare gas atoms stabilizing the cavity have no influence on the static EFG and move too fast to create any time dependent effects .

A strong support for this new model comes also from a comparison of the cavity induced EFGs with the EFGs observed for isolated ^{111}In probes in terrace sites of external Cu, Ni and Au surfaces as investigated by Schatz and coworkers [13]. In all cases where the EFG was measured in equally oriented epitaxially grown thin films the results were in very good agreement with our values. For the comparison in some cases one has to correct for the effect that the EFGs measured in the surface of bulk single crystals are about 10% smaller than in the corresponding thin films.

The decrease of the fraction of ^{111}In probes in this configuration (Fig.1) has to be ascribed to the thermally activated migration of the probes into the bulk, where in a cubic environment no EFG is present. This and not the annealing of the cavity which according to TEM investigations in all metals occurs at much higher temperatures is responsible for the disappearance of the characteristic PAC signal.

As described above in the bcc metals quadrupole interaction frequencies with the same characteristics can be observed not only after rare gas implantation but also in virgin samples. However the experimental evidence also favours the model of cavity formation at the PAC probe in Mo and W. First it is known [14, 15] that ^{111}In traps vacancies in annealing stage three of these metals. The defects in question appear just at the temperature where the clustering of vacancies is expected. In this temperature range the appearance of cavities in the absence of rare gas impurities has also been observed in TEM investigations [16, 17]. Theoretical calculations also come to the result that in the bcc metals the formation of threedimensional vacancy clusters is energetically favourable in contrast to the fcc metals [11]. Therefore the Indium probe atoms seem only to serve as nucleation centers for the initial stage of the cavity formation.

Further it is interesting to note that the ratio between the quadrupole interaction frequencies induced by a single vacancy to that observed in a surface is ~0.5 in all previously discussed fcc metals. The fact that the same ratio is observed for the bcc metals can be taken as additional evidence that the ^{111}In probe indeed occupies a surface site.

Finally the results of the PAC studies are in good agreement with TEM experiments on the cavity growth in the different metals. Under similar conditions at temperatures usually above the range where PAC first detects cavities the formation of large cavities is observed. In the fcc metals it has also been found that the bounding planes of the cavities are predominantly the same crystallographic planes as determined by PAC i.e. $\langle 111 \rangle$ and $\langle 100 \rangle$.

b. Temperature dependence of the surface EFG

The electric field gradient in noncubic metals generally has been found to vary with the temperature according to the $T^{3/2}$ rule [18]. For the surface layer of a metal it is expected [19] that due to the quasi-two dimensional phonon spectrum the EFG should depend differently on the temperature and as T approaches 0 K should even become constant. Indeed Platzer et al. [20] have found a linear dependence of V_{zz} for ^{111}In in external Cu (111) surfaces between 77 and 600 K:

V_{zz} (T) $= V_{zz}$ (0) (1 - B T) with B $= 11.1(3) \times 10^{-5}$ K^{-1}.

In this experiments the temperature range below 77 K was not accessible due to the condensation of residual gas on the unprotected external surface. This problem does not occur ihn the case of inner surfaces and therefore we were able to carry out a precise study of the temperature dependence of the EFG between ~ 150 K and 12 K. The result of the measurement is shown in figure 3. Included are also the results of Pleiter et. al. [10] obtained with a He implanted sample. As indicated by the straight line the data can be very well described by a linear temperature dependence over the whole temperature range. From a fit of the

Table 2:

Copper	$\omega = 212(3)$ Mrad/s [1]	$\eta \le 0.2$	$V_{zz} \parallel \langle 111 \rangle$
Nickel	$\omega = 225(5)$ Mrad/s	$\eta \approx 0.15$	$V_{zz} \parallel \langle 111 \rangle$ [2]
	$\omega_L = 43(4)$ Mrad/s [3]		
Gold	$\omega = 212(2)$ Mrad/s	$\eta = 0.$	
Aluminum	$\omega = 138 - 128$ Mrad/s [4]	$\eta = 0.$	$V_{zz} \parallel \langle 100 \rangle$ [5]
Molybdenum	$\omega = 212(2)$ Mrad/s	$\eta \approx 0.15$	$V_{zz} \parallel \langle 100 \rangle$
Tungsten	$\omega = 248(3)$ Mrad/s	$\eta = 0.$	$V_{zz} \parallel \langle 100 \rangle$

[1] after annealing at 523 K
[2] Orientation taken from ref. [6]
[3] Larmor precession frequency at $T_m = 500$ K
[4] after annealing between 293 – 773 K
[5] Orientation taken from ref. [7]

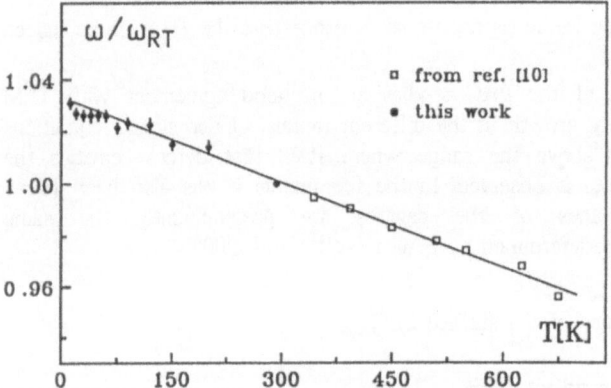

Figure 3: Temperature dependence of the cavity induced quadrupole precession frequency ω normalized to the room temperature value ω_{RT} for the host Copper.

relation given above one obtains a value $B = 10.9(3) \times 10^{-5}$ K^{-1} in excellent agreement with the value for external surfaces. However it is interesting to note that all data points below 75 K lie below the fitted line. This could be taken as an indication of the above mentioned irregular behaviour at low temperatures but the statistical significance is considered to be insufficient to finally decide the question.

The characteristic quadrupole precession frequencies observed in Mo and W (Tab. 2) also show a linear temperature dependence (Fig.4). The parameter B has very similar values namely $9.6(16) \times 10^{-5}$ K^{-1} in Mo and $10.3(10) \times 10^{-5}$ K^{-1} in W. This can be taken as further evidence that the assignment of these frequencies to In probes in the inner surfaces of cavities in these metals has been correct.

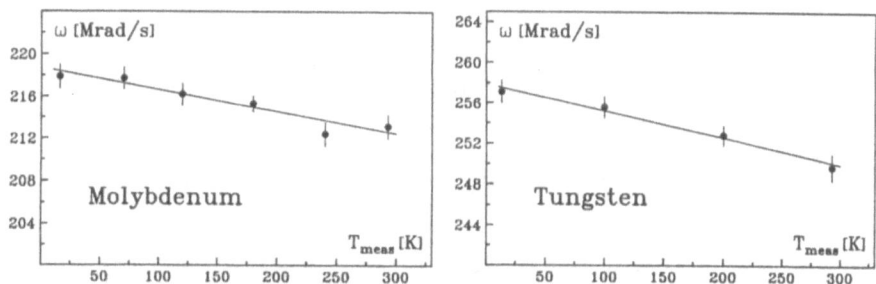

Figure 4: Temperature dependence of the cavity induced quadrupole precession frequency ω for Molybdenum and Tungsten.

c. Growth mechanism for cavities in Cu

As discussed above the appearance of the characteristic precession frequency due to the formation of cavities indicates the simultaneous trapping of vacancies and rare gas atoms at the PAC probe. This implies that both species must be mobile in the same temperature range. In the first experiments in Cu however it was already observed that there seemed to be a certain correlation between the onset of the cavity formation and the rare gas species involved [4]. Since in these experiments disturbing influences due to different probe preparation procedures could not be excluded a systematic study of the growth mechanism of the cavities was performed. To this end the cavity formation probability was measured in an isochronal annealing program with three identical samples implanted with Ne, Kr and Xe, respectively. To exclude any differences in annealing temperature all annealing steps were carried out simultaneously in the same furnace.

As can be seen in figure 5a the fraction of ^{111}In probe atoms experiencing the surface EFG in cavities indeed increases at lower temperatures in the Ne implanted sample than in the ones containing Kr and Xe. More pronounced this effect can be seen in the differentiated curve (fig. 5b) where the steepest increase is found for Ne at ~450 K whereas it occurs only at ~600 K for Kr and Xe. This

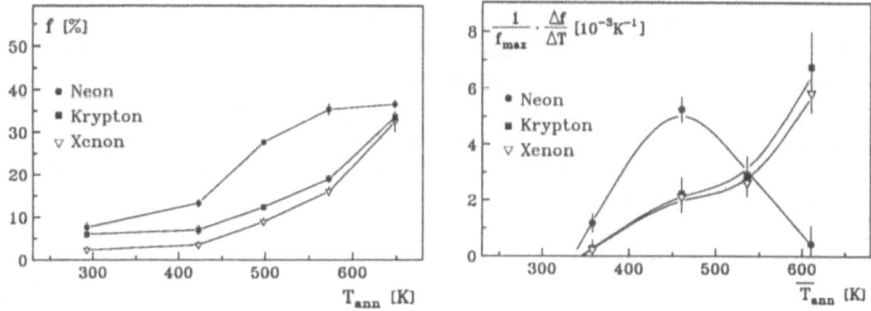

Figure 5: Variation of the fraction of ^{111}In probe atoms in cavities versus annealing temperature (left). On the right the differential variation of the same data is shown.

is surprising since the implantation of the heavier atoms produces a considerably higher number of vacancies in the Cu lattice [21].

Since the onset of vacancy migration is only a property of the Cu lattice one has to conclude from this behaviour that the cavity formation is initiated by the onset of the rare gas migration which occurs at lower temperatures for Ne than for Kr and Xe. The following description of the process can then be derived from the data.

After the implantation at 293 K all rare gas atoms occupy substitutional sites since it is known that they either are very mobile interstitials (He) [9] and are trapped in vacancies, i.e. become substitutional in very short times or do not form stable interstitials at all and immediately occupy a substitutional site (Kr, Xe) by pushing a lattice atom off its site [22].

The diffusion of the heavy rare gases is therefore expected to occur via the conventional vacancy mechanism with an activation energy E_a not very different from the self diffusion energy:

$$E_a = E_V^F - E_{RGV,V}^B + E_{RGV,V}^M.$$

Here E_V^F is the formation energy of a vacancy, $E_{RGV,V}^M$ the migration enery of the complex and $E_{RGV,V}^B$ the binding energy between a substitutional rare gas atom and a single vacancy. The values for the last quantities are not known. However if one assumes that the rare gas atom behaves exactly like a Cu lattice atom E_{RGVV}^M is equal to the migration enery of a vacancy and $E_{RG,V}^B$ vanishes. Then E_a is 2.2 eV [23]. If an attractive interaction exists between a substitutional rare gas atom and a vacancy this leads to a reduction of E_a, i.e. a faster diffusion. This seems to be the mechanism most probable for the diffusion of Kr and Xe. However since In diffuses via the same mechanism it cannot be excluded that the increase of probe atoms in cavities around 600 K is due to the trapping of In at already existing cavities.

In contrast the mobility of Ne at much lower temperatures cannot be understood in this picture. Especially since one would expect in the same scheme a similar or even higher activation energy since th binding energy of Ne to a vacancy should be smaller than for Kr and Xe. Therefore the migration of Ne at lower temperatures can only be explained by a different mechanism. For He Jung et al. [24] have suggested that the "dissociative mechanism" is responsible for the migration. In this case the rare gas atom dissociates from its substitutional site and migrates interstititally i.e. very fast until it is trapped at a probe atom - vacancy complex. The activation energy is then determined by the binding energy of the gas atom to the vacancy or the vacancy complex it stems from. This can be described by the reaction:

$$RG_m V_n \rightarrow RG^{int} + RG_{m-1} V_n$$

with $RG_m V_n$ and $RG_{m-1} V_n$ denoting rare gas - vacancy complexes consisting of m and n partners respectively and RG^{int} is an interstital rare gas atom.

This process has been invoked to explain the results of thermal desorption (THDS) experiments of He in Cu [24] and the early mobility of Ne observed in the present experiments also confirms its validity.

For the heavier rare gases this process cannot take place since theoretical calculations have shown [22] that due to their large size if placed in an interstitial position they immediately push a lattice atom off its site and thus become substitutional.

4. Summary

PAC measurements have shown that In impurities act as nucleation centers for cavity formation in several fcc (Al, Ni, Cu, Au) and bcc (Mo, W) metals. In the fcc metals the presence of implanted rare gases is a necessary condition for the formation of threedimensional vacancy clusters whereas in Mo and W they also form in pure samples. From the properties of the measured EFG it was concluded that the In probe atoms are incorporated in the inner surfaces of the cavities. The EFG showed a linear temperature dependence as expected from theoretical calculations. Concerning the migration mechanism for rare gases in Cu it was found that the heavy rare gases diffuse via the usual vacancy mechanism whereas for the light gases He and Ne the so called "dissociative mechanism" proposed in the literature [24] could be verified.

5. Acknowledgement

The autors are indebted to Drs. W. Kesternich and H. Ullmaier for the accompanying TEM investigations and helpful discussions. This work was partially supported by the Bundesminister for Science and Technology under contract no: 03-B02BON.

6. References

[1] H.J. von den Driesch and P. Jung, High Temperature High Pressure 12 (1980) 635

[2] R. Vianden in "Nuclear Physics Applications on Materials Science", eds. E. Recknagel, J.C. Soares, Kluwer Academic Publishers, Dordrecht (1980) 239

[3] Proceedings of the workshop on "Submicroscopic investigations of defects in semiconductors", Elsevier/North Holland, Amstedam (1991) in press

[4] R. Schumacher and R. Vianden, Phys. Rev. B 36 (1987) 8258

[5] R. Schumacher, Dissertation, University of Bonn (1991)

[6] F. Pleiter, A.R. Arends and H. de Waard, Phys. Lett. 77A (1980) 81

[7] F. Pleiter, K. Post, M. Mohsen and T.S. Wierenga, Phys. Lett. 101A (1984) 363

[8] M. Deicher, G. Grübel, W. Reiner, E. Recknagel and Th. Wichert, Annual Report University of Konstanz (1984) p. 11

[9] G. Grübel, Dissertation, University of Konstanz (1987)

[10] F. Pleiter in "Solid State Reactions After Ion Implantation", eds. K.P. Lieb and M. Uhrmacher, University of Göttingen (1986) 125

[11] S.J. Zinkle, L.E. Seitzmann and W.G. Wolfer, Phil. Mag. 55 (1987) 111

[12] O. Echt, E. Recknagel, A. Weidinger and Th. Wichert, Z. Phys. B 32 (1978) 59

[13] J. Voigt, Dissertation, University of Konstanz (1990)

[14] A. Weidinger, R. Wessner, E. Recknagel and Th. Wichert,
Nucl. Instr. Meth. 182/183 (1981) 509

[15] U. Pütz, A. Hoffmann, H.J. Rudolph and R. Vianden, Z. Phys. B 46 (1982) 107

[16] M. Eldrup, O.E. Mogensen and J.H. Evans in "Proceedings of an International
Conference on Fundamental Aspects of Radiation Damage in Metals", Vol. II, eds.
M.T. Robinson and F.W. Young, Gatlinburg (1975) 1127

[17] K.D. Rasch, R.W. Siegel and H. Schultz, Phil. Mag. A 41 (1980) 91

[18] R. Vianden, Hyp. Int. 15/16 (1983) 189

[19] D.R. Torgeson and F. Borsa, Phys. Rev. Lett. 37 (1976) 956

[20] R. Platzer, R. Fink, T. Klas, G. Krausch, J. Voigt, R. Wesche and G. Schatz,
Annual report, University of Konstanz (1987) p. 61

[21] J.P. Biersack and L.G. Haggmark, Nucl. Instr. Meth. 174 (1980) 257

[22] M.I. Baskes, C.L. Bisson and W.D. Wilson, J. Nucl. Mater. 83 (1979) 139

[23] N.L. Peterson, J. Nucl. Mater. 69/70 (1978) 3

[24] P. Jung and K. Schroeder, J. Nucl. Mater. 155–157 (1988) 1137

Plasma-Wall Interaction
in Controlled Thermonuclear Fusion Research

R. Behrisch

Max-Planck-Institut für Plasmaphysik, EURATOM Association,
W-8046 Garching bei München, Fed. Rep. of Germany

For about 30 years, there have been worldwide efforts to
demonstrate the possibility of gaining energy by controlled
thermonuclear fusion of hydrogen atoms on earth, in a plasma with a
temperature of a few 100 million degrees (corresponding to about
12 keV). It is tried to confine the plasma in a vessel by closed
nested magnetic surfaces. In such experiments the interaction of the
plasma with the vessel walls constitutes a very critical problem. The
plasma-wall interaction is a consequence of the limited particle and
energy confinement in a magnetic field. This limited confinement is
necessary, however, to be able to exhaust the He ash produced and the
energy desposited in the plasma. With a continuously burning D,T
fusion plasma, the particle and energy fluxes which have to leave the
plasma are given by the production rates of He by fusion reactions. In
a fusion device the vessel walls act as a sink and source for ions
from the plasma. These processes may lead to uncontrolled density
changes in the plasma, and they represent a potential tritium
inventory problem. Further the vessel walls are a source and sink for
atoms of the vessel wall material. These constitute impurities in the
plasma which cool the plasma by radiation and dilute the hydrogen
density for a given electron density. Ion implantation and deposition
as well as sputter erosion by plasma particles and local deposition of
high power on the vessel walls cause critical destructions of the
vessel wall material. Plasma-Wall Interaction phenomena may possibly
be mitigated if the plasma in front of the vessel walls can be
operated at a high density and a very low temperature, i.e. well below
the threshold energy for sputtering, if the heat load from the central
plasma can be uniformly distributed on the vessel walls, and if an
appropriate low-Z material is selected for the plasma-facing areas of
the vessel walls.

1. Introduction

The topic of plasma-wall interaction or plasma-solid interaction (PSI)
deals with the interaction of matter in two extreme states, i.e. the
physics of the processes occurring when a hot plasma comes into direct
contact with a solid, generally without the liquid and gaseous states

in between. To understand the physics of this situation, detailed
information about different fundamental processes is needed, such as
the interaction of energetic particles and large power fluxes with
solids. During the interaction with the plasma the surface layers of
the solid are modified by the depositions from the plasma and by
erosion, and the plasma is modified by the loss of particles and
energy and by the introduction of atoms from the vessel walls, which
in part constitute impurities in the plasma. The most powerful
technique for investigating the modifications of the plasma exposed
surface layers of solid walls has been found to be analyis with high-
energy (MeV) ion beams. These processes are central topics of this
symposium.

Plasma-solid interaction is a disturbing process in fusion research
[1], but it is also widely applied in present high technology, such as
controlled surface erosion as well as ion implantation and deposition
of layers and protective coatings on surfaces [2].

2. Nuclear Fusion

Nuclear fusion of light atoms is a potential energy source. This is
demonstrated by the sun and it was shown by the unfortunate explosion
of the first hydrogen bomb on earth in 1952. The question is: "How can
we tame and control it in the environmental conditions on our planet?"

The fusion reaction with the largest cross-section at the lowest
energy is the D,T reaction:

$$D + T \quad = \quad {}^{4}He \ (3.5 \ MeV) + n \ (14.1 \ MeV)$$

$$(10 \ to \ 100 \ keV) \qquad (17.6 \ MeV)$$

The amplification factor of this reaction is about 400. Only this
fusion reaction is being regarded for a first fusion reactor.
Reserach on controlled thermonuclear fusion started in the 1950s.
Since the Second International Conference on "Peaceful Uses of Atomic
Energy" in Geneva in 1958 [3], there has been a worldwide
collaboration in this field and considerable progress has been
achieved in the last 30 years [4]. However, the scientific goal, that
is, to build a first energy-producing fusion device, which would
release more energy than is needed for initiating fusion processes and
keeping the device operating, has not yet been reached.

In the years 1989/1990 an international team of 50 experts from
Europe, Japan, the Soviet Union, and the United States hosted by Max-
Planck-Institut für Plasmaphysik in Garching, Germany, has worked out
a conceptional design study for a first nearly continuously burning
fusion reactor, named ITER (International Thermonuclear Experimental

Reactor) [5,6]. One of the most serious problems in this design
proposal is "particle and power exhaust", which is closely related to
the *plasma-material interaction* [1,6].

3. The Fusion Plasma, Magnetic Confinement

The most favourable condition for getting a sufficient number of
fusion reactions for a positive energy balance is to heat the D,T gas
to temperatures of 10 to 20 keV. At such temperatures matter is in the
plasma state, i.e. a thermonuclear plasma. Such a plasma cannot be
confined in a vessel by the solid walls for any extended time. The
vessel walls would melt and sublimate. However, a plasma can be
confined for some time by a magnetic field, such as in closed, nested
magnetic surfaces (Fig. 1) inside a vacuum vessel, which protects the
plasma from the outer atmosphere. Most promising is a toroidal
geometry, such as applied in the tokamak or stellarator [7]. The
confinement of a plasma in a magnetic field is limited. The electrons
and ions gyrate and move freely along the magnetic field lines, which
form the magnetic surfaces, and they also diffuse and drift
perpendicularly to the magnetic surfaces. This means that the plasma
expands up to the last closed magnetic surface, namely the separatrix,
and into an area where the magnetic surfaces are no longer closed. The
last closed magnetic surface may be given by a limiter (Fig.1). It may
also be determined by the structure of the magnetic field which is

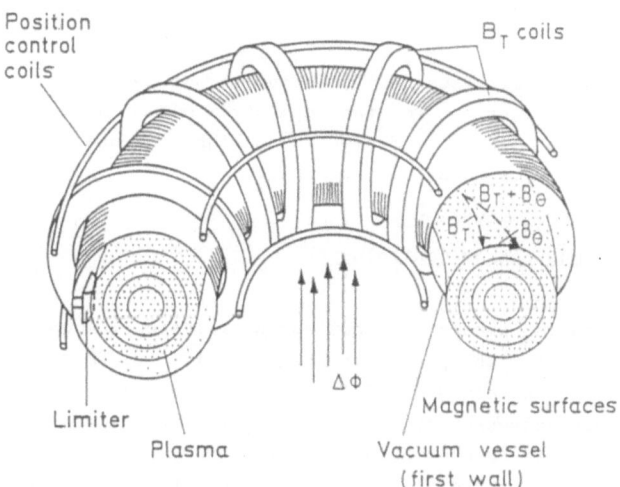

Figure 1: Schematic of a limiter Tokamak. The plasma is confined in
the vacuum vessel by closed nested magnetic surfaces, which are
produced by external coils and by the toroidal current induced in the
plasma by a changing magnetic flux in the center. The last closed
magnetic surface, i.e. the separatrix, is given by a limiter.

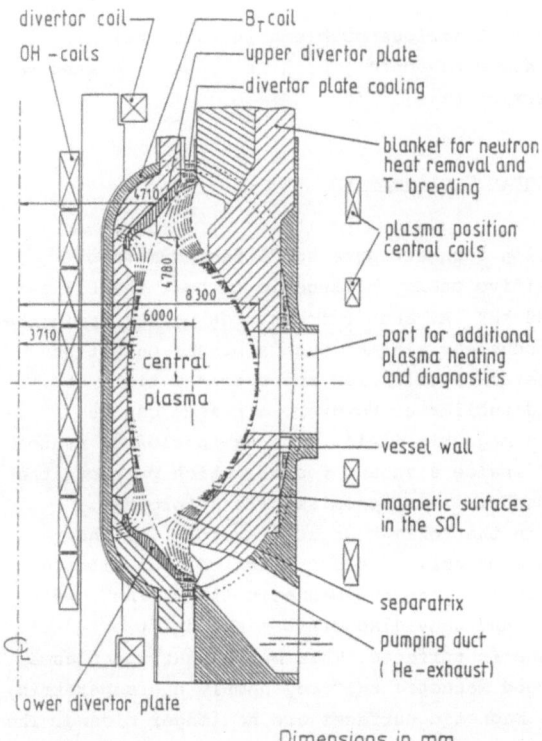

divertor coil
OH –coils
Bᵣcoil
upper divertor plate
divertor plate cooling
blanket for neutron heat removal and T-breeding
plasma position central coils
port for additional plasma heating and diagnostics
vessel wall
magnetic surfaces in the SOL
separatrix
pumping duct (He-exhaust)
central plasma
lower divertor plate

Dimensions in mm

Figure 2: A vertical cut of the planned ITER (International Thermonuclear Experimental Reator). The separatrix is determined by external divertor coils. The magnetic surfaces outside the separatrix intersect the top and bottom divertor plates.

produced by the external coils, such as in a divertor geometry (Fig.2). The magnetic surfaces outside the separatrix end on areas of the vessel wall which are of special design for a high heat and particle load, such as limiters and divertor plates.

4. Necessary Particle and Energy Confinement

To produce a burning fusion plasma with net energy release, the energy and particle confinement for the plasma in the magnetic field must be long enough to reach and sustain a sufficient plasma temperature and density. This was first expressed by the LAWSON criterion which requires that for a positive energy balance it is necessary in addition to achieve a temperature of $kT \geq 12$ keV, that the condition $n_e \tau_E \geq 3 \ 10^{14}$ scm^{-3} be met [8], where n_e is the electron density in the plasma and τ_E the energy confinement time. To achieve continuous burn of a fusion plasma, a similar criterion has to be satisfied, i.e. the energy deposited in the plasma by the 3.5 MeV α-particles produced

in the fusion reactions and any additional injected power for heating
the plasma must compensate the energy losses from the plasma due to
radiation and conduction [9].

The particle and energy confinement in the fusion plasma must,
however, also be limited in order to exhaust the ^{4}He and the α-energy
deposited in the plasma i.e. to prevent poisoning by the ash and
overheating of the plasma [10-15]. From the viewpoint of the necessary
particle and energy exhaust the limited confinement of the
thermonuclear plasma in a magnetic field is advantageous. However, it
must be possible to adjust and control both the energy and particle
confinement in accordance with the fusion reaction rate [15]. The
major effort in fusion research in the past decades has been to
improve the energy confinement to reach the temperatures and the
values for $n_e \tau_E$ which are needed for a fusion plasma to ignite
[4,7,8]. The necessary exhaust of the He ash has only recently become
of concern [10-15].

The limited plasma confinement and the necessary exhaust of particles
and energy for a continuously burning fusion plasma cause fluxes of
energy and particles to the vessel walls which must be equal to those
given by the fusion reaction rate. The values for these fluxes which
are given in the following refer to the ITER design [6,16], they are
similar, however, to those predicted for several fusion reactor design
proposals with are planned for the same nuclear power.

5. He and Energy Production Rates for a Burning D,T Plasma

The burn of a thermonuclear D,T plasma is described by means of the
fusion reaction rate $<\sigma v>$, which is obtained by multiplying the fusion
cross-section 'σ' by the relative particle velocity 'v' and averaging
with the velocity distribution of the ions, which is mostly assumed to
be Maxwellian. For a D,T plasma at 10 keV the reaction rate is
$<\sigma v> = 10^{-16} cm^{-3} sec^{-1}$ [7] and for a plasma density of $n_D = n_T = 10^{14}/cm^3$
[5,6] the He and energy production rates per unit volume in the plasma
are given by

$$dn/dt = n_D n_T <\sigma v> = 10^{12} \ ^4He/cm^3 s$$

$$dE/dt = n_D n_T <\sigma v> 17.6 \ MeV = 3 \ W/cm^3$$

In a first fusion reactor, such as ITER, the volume of the reacting
toroidal plasma is planned to be about 350 m^3 and the surface area
about 800 m^2 giving a total nuclear power of 1 GW (200 MW α-power and
800 MW neutrons), a ^{4}He production rate of 3.5 10^{20}/s and a 14.1 MeV
neutron wall flux of 1 MW/m^2 corresponding to about 4.5 10^{17}/m^{2}s.

This α-power and the He produced must leak out of the magnetic
confinement across the separatrix, together with D and T ions. They
then have to be exhausted at the vessel walls.

6. Particle and Energy Fluxes to the Different Areas of the Vessel Walls

For a magnetically confined plasma the particle and energy fluxes
leaving the plasma are expected to be concentrated predominantly on
the limiters (Fig.1) and divertor plates (Fig.2) around the
intersection of the separatrix, i.e. on typically a few % of the
vessel wall area. The specific plasma load on these areas becomes
comparable to or even larger than the average load on all vessel wall
components for a plasma without magnetic confinement. The energy lost
from the plasma by radiation (presumably about 50 % of the α-power)
will be distributed uniformly on all areas of the vessel wall. They
will be further bombarded by a flux of neutral atoms with energies
corresponding to the plasma temperature. These neutrals are created in
charge exchange (CX) collisions between neutrals entering the plasma
from the vessel walls and plasma ions. The vessel walls will be
finally bombarded by the 14 MeV neutrons from the fusion reaction and
the lower energy neutron flux initiated in the vessel wall structure
by the 14 MeV neutrons, which are not confined by the magnetic field.
The intensities of all these energy and particle fluxes as expected
for a fusion reactor with a nuclear power of about 1 GW are summarised
in table 1 [6,15,16].

Table 1 Expected fluxes to different areas of the vessel walls for a
fusion reactor such as ITER with a nuclear power of 1 GW, a vessel wall
area of 800 m^2 and an effective divertor area of about 5m^2. Further, a
recycling factor of about 10^4 is assumed to reduce the average
particle energy to the 100 eV range [15,16].

	Energy fluxes (peak)				Particle fluxes (peak)			
	14 MeV neutrons (MW/m^2)	electrom. radiation (MW/m^2)	He,D,T ions (MW/m^2)	D,T(CX) neutrals (MW/m^2)	14 MeV neutrons ($m^{-2}s^{-1}$)	D,T(CX) neutrals ($m^{-2}s^{-1}$)	He ions ($m^{-2}s^{-1}$)	D,T ions ($m^{-2}s^{-1}$)
vessel wall	1	0.125	-	0.03	$4.5\ 10^{17}$	$5\ 10^{18}$	-	-
div. plate	1	0.125	8 (20 to 40)		$4.5\ 10^{17}$	$5\ 10^{18}$	$3.5\ 10^{22}$	$3.5\ 10^{23}$

7. Plasma-Solid Interaction Processes

If a plasma comes in contact with a solid an electrical potential
builds up, the plasma getting positively and the solid negatively
charged. The electric fields may initiate electrical arcs between the

plasma and the vessel walls. As a consequence of the particle and
energy fluxes, several processes are initiated at the vessel walls,
such as ion backscattering, implantation, trapping, diffusion and
release, sputtering, local heating, melting, and sublimation of the
surface layers.

7.1 The Langmuir Sheath Potential

In a plasma the electrons have generally a much larger velocity than
the ions. The surface of a solid exposed to the plasma is hit by more
electrons than ions. An electrical potential builds up so that low
energy electrons are reflected and only high energy electrons reach
the surface at the same rate as ions. The potential drop, which
amounts 3 kT_e for a hydrogen plasma, occurs in a thin layer in front
of the surface, named the Langmuir sheath. The ions are accelerated in
the Langmuir sheath and they hit the surface with an average energy of
about 5 kT [1].
The electrical potential between the plasma and the solid may be
increased during unstable plasma operation. Electrical arcs, also
named unipolar arcs, may ignite. The vessel walls represent the
cathode and the plasma represents the anode [1]. The electrical arcs
cause an additional erosion of the vessel walls mostly in the form of
small clusters and for metals in the form of small droplets. In
today's tokamaks traces of such electrical arcs are found
predominantly on those areas of the vessel walls which have been
exposed close to the separatrix. Further small metal droplets are
found on the vessel walls around those areas made of metal, where
arcing was observed.

7.2 Ion Backscattering, Implantation, Trapping, Diffusion and Release

The hydrogen ions and atoms bombarding the vessel walls from the
plasma with energies in the 10 eV to keV range are partly
backscattered into the plasma in collisions with atoms in the surface
layers of the solid. The ions which are not backscattered are slowed
down and come to rest in the lattice of the solid.

Backscattering of hydrogen ions from the surface layers of different
solids has been investigated in some detail, both experimentally as
well as by computer simulation for the parameters of interest in
fusion research [17-19]. For these energies typically between 20 % and
50 % of the incident ions are backscattered, mostly as neutral atoms.
The backscattering yields increase with decreasing energy of the ions
and for increasing mass of the solid [17-19], Fig.3. In plasma

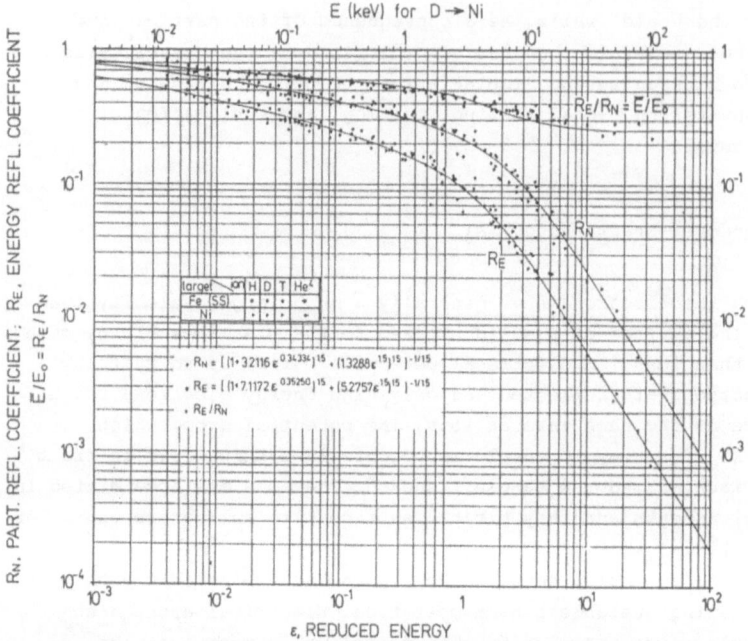

Figure 3: Particle and energy reflection coefficients for light ions incident on Fe and Ni targets. The reduced energy is given by $(M_1/(M_1 + M_2))(a/Z_1 Z_2 e^2)E$, with Z_1, Z_2, M_1, M_2 being the atomic numbers and masses of the incident ions and target atoms and a the Thomas-Fermi screening length [17-19].

experiments the neutrals backscattered from the vessel walls enter again the plasma where they are ionised.

The ions coming to rest in the solid may be permanently trapped, or, depending on the temperature they may diffuse back to the plasma side or into the bulk. The hydrogen ions permanently trapped and those diffusing into the bulk are lost from the plasma. The vessel walls pump initially [20, 21, 22] and the hydrogen atoms lost are replaced by refuelling such as by gas blow or by injection of solid frozen hydrogen pellets. In today's plasma experiments the plasma density can be well controlled in this way.

This scheme for density control is applicable only if the hydrogen atoms implanted into the vessel walls during a discharge are released between discharges such as observed for most metal walls. If the hydrogen ions implanted in the vessel walls are not released between discharges, such as for carbon walls, they accumulates up to saturation. In subsequent plasma discharges they may be released by the plasma load and cause an uncontrolled increase of the plasma density [23]. As a consequence of the plasma induced release, the ions in the plasma are predominantly those of the previous discharges.

Further, the hydrogen trapped in the vessel walls represents a
critical tritium inventory problem in a future fusion reactor. In
order to reduce this problem of density control and hydrogen
inventory, the plasma temperature in front of the vessel wall plates
should be very low, i.e. in the 10 eV range and the target wall
material should not trap implanted hydrogen permanently.

7.3 Sputtering, Release of Wall Atoms

The bombardment of the vessel walls with energetic ions from the
plasma causes a release of wall atoms due to sputtering. Sputtering
yields, i.e. the average number of atoms released at the surface of a
solid due to the impact of one energetic ion are shown in Fig. 4 for
different vessel wall, limiter and divertor plate materials [24].
Materials of low-Z elements especially different forms of carbon are
favoured for the plasma-facing vessel wall areas. This is because for
low-Z elements much larger concentrations can be tolerated in the
plasma, than for high-Z elements.

The sputtering yields of solids depend largely on the energy of the
incident ions. Below a threshold energy, in the 10 eV range no
sputtering is found. At energies above the threshold energy the yields
strongly increase and they have a maximum in the range of keV
energies, see Fig.4, [24].

In fusion research sputtering is especially critical at the divertor
plates and the limiters, because of the large ion fluxes at these
areas. For a burning fusion reactor these fluxes may cause an erosion
of up to 1 cm per day [16]. Two schemes are followed to reduce the
impurity introduction from these areas, i.e. to reduce the erosion and

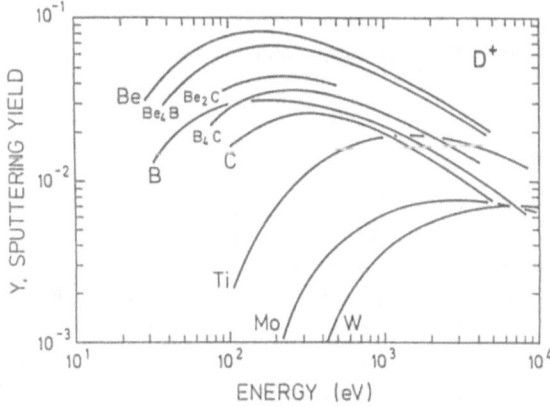

Figure 4: Sputtering yields for different solids at bombardment with
D^+ ions at normal incidence and different energies.

33

to prevent the sputtered particles from entering the central plasma. This may be achieved if a high density, low temperature plasma can be built up in front of the limiters and divertor plates, i.e. shifting the temperature and density gradients between the plaasma and the solid partly into the plasma. The plasma temperature has to be low enough so that the wall bombardment takes place at erergies below the threshold energy for sputtering. Further, the plasma density has to be high enough, so that any sputtered atoms are ionised close to the vessel wall and be pushed back onto the surface where they will be redeposited. The low temperature, high density plasma is expected to be obtained by high recycling of the hydrogen ions bombarding the surface.

There is further sputtering at all areas of the vessel walls due to the bombardment with charge exchange neutral atoms, which have energies corresponding to the central plasma temperature. The flux of energetic neutrals per unit area of the vessel walls is much smaller than the ion fluxes on the limiters and divertor plates. However due to the large area of the vessel walls the total impurity release due to sputtering at these areas may become very critical for a fusion reactor.

7.4 Thermal effects at the vessel walls.

Another critical problem in plasma wall interaction is the large energy deposition from a thermonuclear plasma, especially the high densities at the divertor plates and the limiters. A material with a large thermal conductivity is needed in order to remove the heat for a continuously burning thermonuclear plasma at a tolerable temperature gradient and plate thickness of the order of 1 cm. In today's large plasma experiments the discharge times are of the order of seconds and inertial cooling contribures. However for such pulsed operation thermal fatigue and cracking is a very serious problem. In the running plasmaexperiments carbon-fiber enforced carbon has shown very promising performance.

7.5 Vessel Wall Modification

The surface layers of the vessel walls of fusion experiments are modified by the hydrogen implantation, by erosion due to sputtering and electrical arcs, the redeposition of eroded material and the energy deposition. Generally at some areas of the vessel wall, such as on limiters and divertor plates close to the intersection of the separatrix, erosion dominates, while at other areas deposition dominates. This causes a transport of wall material. At deposition dominated areas the built up of layers in the range of several μm are observed in today's plasmaexperiments. These layers, which have also been named "tokamakium", generally consist of all materials exposed to the plasma. The layers are mostly implanted with hydrogen up to saturation and they sometimes tend to flake. The surface layers of the vessel wall are also modified by the thermal load. Especially at the divertor plates close to the intersection of the separatrix and at other areas, acting as limiters, melted areas are sometimes found, and surface cracks are observed. Selecting an appropriate material for the vessel walls of a future fusion reactor, which can tolerate these effects without beeing destroyed, will be a mayor task for building a controlled thermonuclear fusion reactor on earth.

References

[1]"Physics of Plasma Wall Interaction in Controlled Thermonuclear Fusion Research", D.E. Post and R. Behrisch (eds.) Plenum Press, New York, London, 1986
[2]"Handbook of Plasma Processing Technology" St.M. Rossnagel, J.J. Cuomo, W.D. Westwood (eds) Noyes Publications, Park Ridge, New Jersey, USA (1989)
[3]Proc. Second United Nations Int. Conf. on the Peaceful Uses of Atomic Energy, 31, 32, United Nations, Geneva, 1958
[4]M. Keilhacker, Phys. Fluids B2 (1990) 1291
[5]K. Tomabechi, Plasma Physics and Controlled Nuclear Fusion Research, 12th International Conference, Nice 1988, NUCLEAR FUSION Supplement, IAEA, Vienna, (1989) 215, and D.E. Post Plasma Physics and Controlled Nuclear Fusion Research, 12th International Conference, Nice 1988, NUCLEAR FUSION, Supplement, IAEA, Vienna, (1989) 233
[6]D.E. Post et al. ITER Physics, ITER Documentation Series, No 21, IAEA Vienna (1991)
[7]F. Engelmann in "Physics of Plasma Wall Interaction", ref [1]
[8]J.D. Lawson, Proc. Phys. Soc. B70, (1957) 6
[9]J.D. Meade, Nucl. Fusion 14 (1974) 289
[10]R. Behrisch, V. Prozesky, Nucl. Fusion 30, (1990) 2166
[11]D. Reiter, G. Wolf, G.H. Klever, Nucl. Fusion 30 (1990) 2141
[12]R.J. Taylor, B.D. Fried, G.J. Morales, Comments Plasma Phys. Contr. Fusion 13 (1990) 227
[13]M.H.Redi, S.A.Cohen, J. Nucl. Mat. 176&177 (1990) 262
[14]M.H.Redi, S.A.Cohen, E.J.Synakowski, Nucl Fusion 31 (1991) 1689
[15]R. Behrisch, NUCLEAR FUSION special supplement on "Plasma Surface Interaction Data for Nuclear Fusion", R. Janev,(ed.) IAEA 1991
[16]M.F.A. Harrison, E.S. Hotston " Two-dimensioal Simulations of the NET/ITER Edge Plasma and Divertor Conditions", NET Report No97, EUR-FU/80/90-97, Mai 1990
[17]R.A. Langley, J. Bohdansky, W. Eckstein, P. Mioduszewski, J. Roth, E. Taglauer, E.W. Thomas, H. Verbeek, K.L. Wilson, NUCLEAR FUSION, SPECIAL ISSUUE, "Data Compendium for Plasma-Surface Interactions", IAEA, Vienna, (1984)
[18]R. Behrisch, W. Eckstein, Ref. [1]
[19]W. Eckstein, NUCLEAR FUSION special supplement on "Plasma Surface Interaction Data for Nuclear Fusion", R. Janev,(ed.) IAEA 1991
[20]W. Poschenrieder, G. Venus and the ASDEX-Team, J.Nucl.Mat. 111/112 (1982) 29
[21]G.M. McCracken, S.J. Fielding, S.K. Erents, A. Propieszyk, P.E. Stott, NUCLEAR FUSION. 18 (1978) 35
[22]TFR Group (P. Platz) J. Nucl. Mat. 93/94 (1980) 173
[23]H. Vernickel, K. Behringer, D. Campbell et. al. J. Nucl. Mat. 111/112 (1982) 317
[24]J. Roth, J. Nucl. Mat. 176/177 (1991) 132

Novel Applications of Narrow Nuclear Resonances*

C. Rolfs**

Ruhr-Universität Bochum, Experimentalphysik III,
W-4630 Bochum, Fed. Rep. of Germany

Abstract: Improvements in targetry and ion beam energy resolution permit the observation of novel effects using narrow nuclear resonances as a probe. For the 400 kV Münster accelerator the ion beam energy resolution was reduced to 15-20 eV at full ion beam current. The development of UHV, vapor, and gas target systems allowed the use of very clean targets with variable density and temperature. With the energy spread of the ion beam approaching the eV level, the dynamics of the nuclear environment (such as atomic shells and solid material) become observable. Some novel applications in nuclear, atomic, molecular, and solid state physics are discussed.

1. Introduction

A wide range of phenomena in pure and applied physics can be studied with narrow nuclear resonances ($\Gamma_R \ll E_R$, Γ_R = few eV). The topics range from the observation of chaos in nuclear spectra to depth profiling measurements. Improvements in the energy resolution of ion beams ($\Delta E/E$) and the quality of targets enable the observation of very weak resonances and improve the sensitivity of applications such as depth profiling. Most importantly for the present discussion, good energy resolution ($\Delta E/E \approx 10^{-5}$) permits investigation of a variety of effects which are not observed with moderate or poor energy resolution ($\Delta E/E \gtrsim 10^{-3}$).

When the energy spread of ion beams (ΔE) approaches the eV level, the dynamics of the nuclear environment - such as atomic shells and solid material - become discernible with narrow nuclear resonances, since the energies in the atomic or solid-state interactions are at the same level. For example, a compound nuclear state is actually a quasistationary state of an entire compound nuclear atom. Therefore, in a nuclear collision it might be possible to observe replicas (or echoes, or satellites) of very sharp nuclear resonances, displaced in energy by the atomic excitation energies. Similarly, the lattice vibrations in a solid target lead to a broadening of a nuclear resonance due to the Doppler motion, which in turn can be used to measure the Debye temperature of the surface layer of the solid.

Here some technical aspects of the work are briefly described, followed by a discussion of results for a few novel applications in pure and applied physics. It should be pointed out that these applications involving the interface between the nucleus and its environment are just beginning and that no comprehensive framework has yet been developed.

2. Ion beam

The energy resolution of ion beams is of central interest at all energies. From various technical considerations it was concluded[1] that a Dynamitron tandem accelerator appears to be ideal. However, the efforts were initially focused on a low energy machine, the 400 kV Münster cascade accelerator. Details of the accelerator including investigations of the ion beam specifications have been described recently[2]. The accelerator provided ion beams of up to 300 μA at the target with an energy spread $\Delta E_I \lessgtr 20$ eV, where the high voltage (HV) ripple was reduced to $\Delta E_{HV} = 4$ eV. The major contribution to ΔE_I arose from the ion source (ΔE_{IS}); recent investigations[3] indicate $\Delta E_{IS} = 15$-20 eV, and thus $\Delta E_I/E \lessgtr 5 \times 10^{-5}$. The energy stability and reproducibility was better than ± 3 eV over a time period of 8 days[2,3]. The accelerator energy scanning (in steps as small as 0.4 eV) and the data acquisition are controlled by a personal computer[3].

3. Target

For high resolution investigations using narrow nuclear resonances, the quality of the target is extremely important. For this reason, an UHV system has been developed which allows in situ fabrication of extremely clean targets of variable thickness. The target temperature T can also be varied. This system[2] was used for studies of solid state effects. For other investigations, such as concurrent atomic and nuclear excitations, windowless gas and vapor targets of variable thickness and temperature have been developed[3-5]. These systems permitted the utilization of extremely thin targets (less than a monolayer, $N_t \lessgtr N_m \cong 1 \times 10^{15}$ atoms/cm^2), with significantly reduced Doppler broadening (in part by employing the crossed beam technique).

4. Concurrent atomic and nuclear excitations

For simplicity, consider a monoenergetic projectile with energy E_p incident on a target atom (monolayer thickness). The projectile excites a narrow nuclear resonance, which is observed via the characteristic nuclear radiation. Since each projectile initiating a resonance must pass through

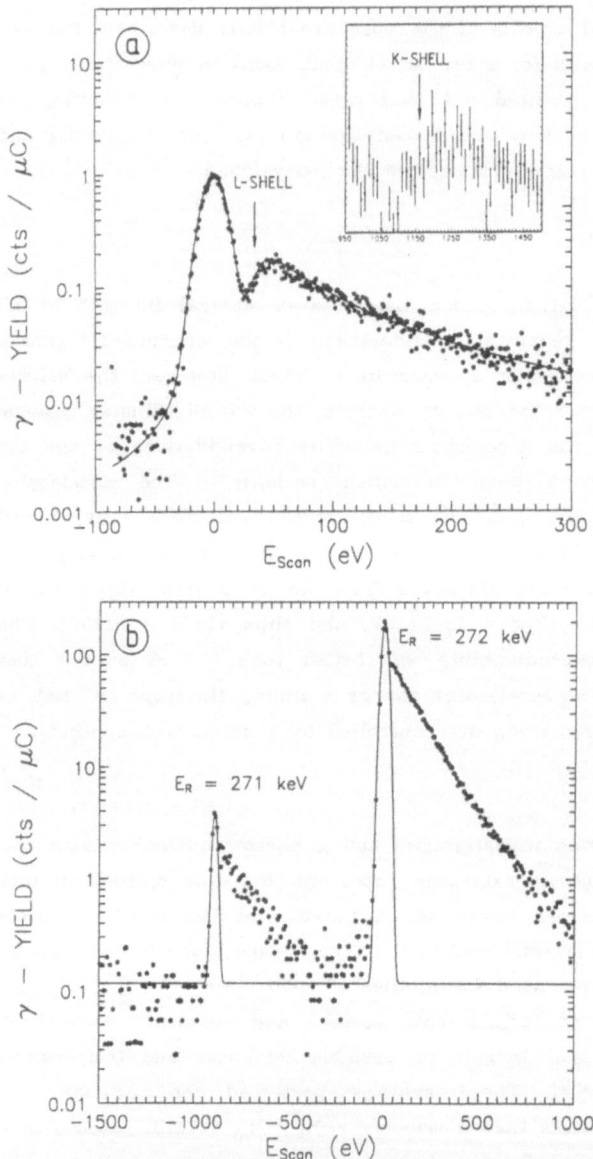

Fig. 1: (a) Yield curve of the E_R = 272 keV resonance in ^{21}Ne(p,γ)^{22}Na obtained at T = 25 K and N_t = 0.3 N_m. The replica resonances correspond to L- and K-shell ionization of the compound atom Na. The curve through the data points are fitted Voigt profiles.
(b) Larger energy scan near the E_R = 272 keV resonance revealing a new resonance at E_R = 271 keV. The Voigt profiles (solid curves) include only the resonance region, i.e. without electron excitations.

the electron cloud of the target atom, it can with some probability excite or knock out an electron. This process results in echoes, or satellites, of the nuclear resonance at higher energies ($E_p > E_R$), corresponding to excitation or emission of K,L,M,... electrons before (or at the same time as) the resonance is formed. The "replica" resonances are displaced in energy by the respective atomic excitation energies, followed by a $1/E^2$ tail in the case of ionization into the continuum. The impact parameter b on the atomic scale is obviously very small and it is usually permissible to assume b = 0. The yield of the replica resonances relative to the total yield reflects the corresponding atomic excitation probabilities $P_{K,L,M}(b=0)$. In real experiments, these structures are broadened due to contributions from the finite energy spread of the ion beam (ΔE_I) and from Doppler broadening of the target atoms (ΔE_D): $\Delta E_{eff} = (\Delta E_I^2 + \Delta E_D^2)^{1/2}$. Therefore, one expects to observe an asymmetric high energy shoulder at $E_p > E_R$, which may still reflect the individual atomic excitations, depending on ΔE_{eff}.

Fig. 1a shows a thin-target yield curve ($N_t = 0.3\ N_m$) obtained at the E_R = 272 keV resonance in $^{21}Ne(p,\gamma)^{22}Na$. One sees clearly an asymmetric high energy shoulder superimposed with edges at E_L = 34±3 eV and E_K = 1150±50 eV, corresponding fairly well to $L_{2,3}$- and K-shell ionization of the compound atom Na, respectively. The curve through the data points is a fit using Voigt profiles, with an $1/E^2$ dependence at energies above the respective edges. The deduced atomic probabilities are $P_{L_2,_3}(b=0)$ = 0.48±0.03 and $P_K(b=0)$ = $(1.0±0.5)\times10^{-3}$, consistent with available atomic physics data. Note that the excitation of the L_1-shell (E_B = 74 eV) is not observed. Similar results also were obtained[3] for other resonances in $^{21}Ne(p,\gamma)^{22}Na$, as well as for the E_R = 309 keV resonance[4] in $^{23}Na(p,\gamma)^{24}Mg$ (here, even M-shell excitation became observable) and the E_R = 184 keV resonance[5] in the $\alpha-\alpha$ scattering (Fig. 2).

Fig. 1b shows data indicating the existence of a nuclear doublet, at E_R = 271 and 272 keV, separated by ΔE_R = 889±5 eV, with total widths of $\Gamma_R \leqslant$ 5 eV and $\leqslant$ 3 eV, respectively. Previous investigations reported only one resonance with $\Gamma_R \leqslant$ 100 eV. This example demonstrates the potential of high resolution ion beams for nuclear physics research, i.e. resolution of doublets (or multiplets) and measurement of resonance widths in the range Γ_R = 1 to 100 eV, which is inaccessible (or difficult) with other techniques.

5. Molecular physics

The E_R = 126 keV resonance in $^{21}Ne(p,\gamma)^{22}Na$ was investigated with a monolayer target and a proton beam (E_p = 126 keV) as well as a H_2 molecular beam (E_{H_2} = 252 keV). The resulting yield curves are shown in Fig. 3: the very broad structure observed with the molecular beam (FWHM =

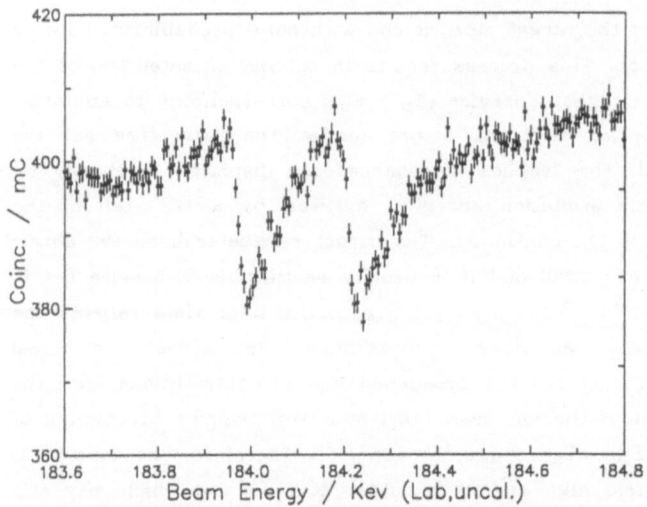

Fig. 2: The relative Mott yield in the α-α scattering at θ_{lab} = 45° is shown
near the E_R = 184 keV resonance (ground state of ^{8}Be), obtained
with an ultra-thin target and the crossed beam technique. The
nuclear resonance shows two replica resonances displaced by about
250 eV, which probably correspond to the $(1s)^2 2s$ and $1s(2s)^2$
atomic configurations of the compound atom Be. The asymmetric
shapes at the high energy edges indicate atomic ionization. The
present results are in fair agreement with earlier work[6], but have
significantly improved precision.

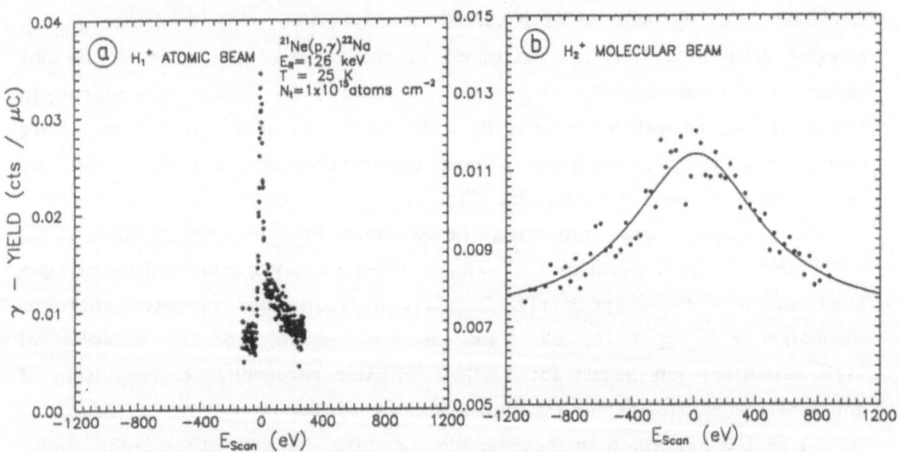

Fig. 3: Yield curve at the E_R = 126 keV resonance obtained with (a) a
proton beam and (b) a H_2 molecular beam. The Lorentzian curve in
(b) has a FWHM of 1000 eV.

1000±100 eV) probably arises from the Coulomb explosion of the molecule. As seen in the figure, a high resolution ion beam amplifies enormously the effects of this process. Such experiments and their analyses may contribute to solutions of some open questions in molecular physics.

6. Energy loss (collision) spectrum

At the projectile energies of the present work, the energy loss of the projectiles in a target is dominated more than 99% by electronic energy loss mechanisms. The projectile (mass m_p) transfers an energy Q to the electrons (mass m_e), which varies between Q_{min} (i.e., the binding energy of the electrons in the target) and Q_{max} (i.e., the classical recoil energy). This energy transfer involves all atomic impact parameters b (both close and distant collisions) and is discrete, varying between Q_{min} and Q_{max} with probability $W(Q)$, called the collision spectrum. For free electrons, $W(Q)$ is described by the Thompson law, $W(Q) \propto Q^{-2}$, and the energy loss per unit path length, dE/dx, is dE/dx $\propto \ln(Q_{max}/Q_{min})$. However, the electrons cannot be considered as free for all values of Q (at least not for Q near Q_{min}) and Q_{min} also cannot be calculated reliably. For these reasons, one takes usually Q_{min} as an adjustable parameter to fit the observed dE/dx value. A detailed knowledge of $W(Q)$ is not only important for the microscopic understanding of stopping power, but also for the Lewis effect (see below).

Measurements of $W(Q)$ were performed in two ways: (i) comparison (subtraction) of yield curves obtained for a target of monolayer thickness with that obtained with a thicker target; and (ii) comparison (subtraction) of yield curves with and without energy loss of the beam, produced in a windowless gas cell (filled with Ne or He gas) prior to the monolayer target. An illustrative example is shown in Fig. 4, where the subtracted yields have already been deconvoluted to correct for energy resolution of the ion beam. One sees an energy loss spectrum with a mean value of 60±20 eV, in fair agreement with stopping power data. The solid curve corresponds to the Thompson law matched at high energy losses. The comparison indicates that the law is not valid at low Q values.

7. The Lewis effect

The quantizied character of energy loss just discussed is manisfested in the Lewis effect on the thick-target yield curve of narrow resonances. The standard signature is a peak (the Lewis peak) near E_R, followed by the usual plateau. Since the charged particles lose energy Q in discrete steps (between Q_{min} and Q_{max}), some particles with an energy $E_p \gtrsim E_R$ will

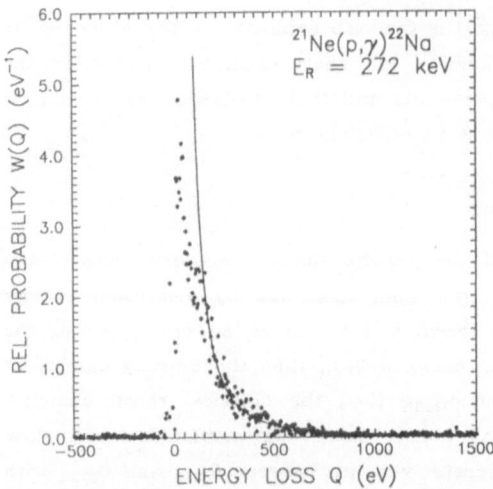

Fig. 4: Collision spectrum obtained from yield curves at the E_R = 272 keV
resonance in $^{21}Ne(p,\gamma)^{22}Na$. The solid curve corresponds to the
$1/Q^2$ Thompson law matched at high Q values.

"jump over" the resonance and do not contribute to the yield; projectiles
incident at $E_p = E_R$, however, all have a chance to react. The thick target
yield is given by the expression

$$Y*(E_p) = \int_0^\Delta Y(E_p-Q) \; W*(Q) \; dQ,$$

where Δ is the target thickness (in units of eV) and $Y(E_p-Q)$ is the
thin-target (monolayer) yield curve including information on Γ_R, ΔE_I, ΔE_D,
and atomic excitations at impact parameter b = 0. The yield curve $Y*(E_p)$ is
thus the integrated convolution of $Y(E_p-Q)$ with the collision spectrum
$W*(Q)$, which can be derived from the single encounter collision spectrum
$W(Q)$. Therefore, $W(Q)$ is the key quantity in the dwescription of the Lewis
effect.

The thick-target yield curve obtained at the E_R = 272 keV resonance in
$^{21}Ne(p,\gamma)^{22}Na$ using a windowless gas target (Fig. 5a) shows clearly the
Lewis peak followed by a few damped oscillations, which reach finally a
plateau. Folding and integrating the measured functions $Y(E_p-Q)$ and $W(Q)$
(Figs. 1a and 4), the resulting curve (simulation) reproduces the data fairly
well. For comparison, data are shown in Fig. 5b at the E_R = 309 keV
resonance in $^{23}Na(p,\gamma)^{24}Mg$, obtained with a solid target (UHV system). The
Lewis peak is also clearly visible but it is smaller in magnitude, mainly due
to the existence of the lattice motion of the solid (zero point vibrations)
leading to a value of ΔE_D = 40 eV. This Doppler width can in turn be used

42

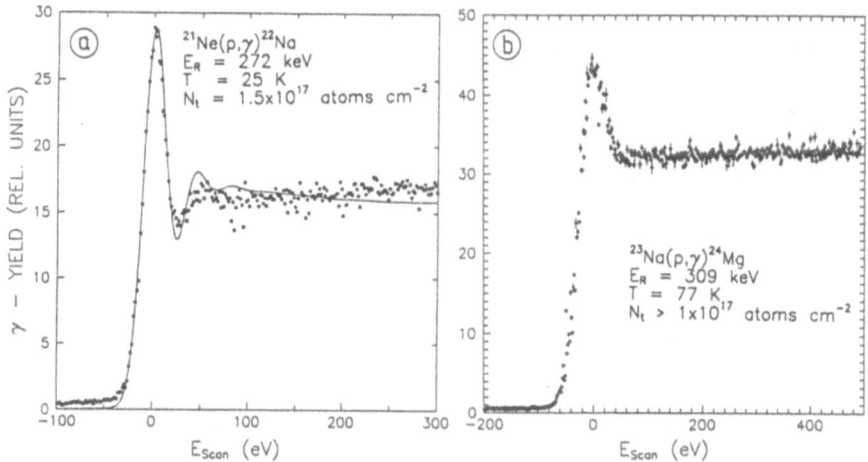

Fig. 5: (a) Thick-target yield curve of the E_R = 272 keV resonance in
^{21}Ne(p,γ)^{22}Na using a windowless gas target (T = 25 K). The
calculated curve through the data points is based solely on
experimental data.
(b) Thick-target yield curve of the E_R = 309 keV resonance in
^{23}Na(p,γ)^{24}Mg using a solid target (T = 77 K).

to extract a value for the Debye temperature of the phonon spectrum at
the target surface. Thus, high resolution ion beams can be used to measure
microscopic solid state properties.

References

1. S. Wüstenbecker et al., Nucl. Instr. Meth. A256 (1987) 9
2. S. Wüstenbecker et al., Nucl. Instr. Meth. A279 (1989) 448
3. W.H. Schulte et al., to be submitted
4. H. Ebbing, Thesis, Universität Bochum (1992)
5. S. Wüstenbecker, Thesis, Universität Bochum (1992)
6. J. Benn et al., Nucl. Phys. A106 (1968) 296

*Work supported in part by Deutsche Forschungsgemeinschaft, US
Department of Energy (Office of High Energy and Nuclear Physics), NATO
Scientific Affairs Division, Minister für Wissenschaft und Forschung des
Landes NRW, NC Board of Science and Technology, and Friedrich-Flick
Förderungsstiftung.
** In collaboration with W.H. Schulte, H. Ebbing, S. Wüstenbecker, H.W.
Becker, M. Berheide and M. Buschmann (Münster/Bochum), G.E. Mitchell
(North Carolina State University, Raleigh, NC, USA) and J.S. Schweitzer
(Schlumberger-Doll Research, Ridgefield, CT, USA).

Molecular Dynamics Simulation
of Cluster-Ion Impacts*

T.A. Tombrello

Division of Physics, Mathematics, and Astronomy,
California Institute of Technology, Pasadena, CA 91125, USA

Bombardment with beams of heavy molecular ions (cluster ions) has
shown initial promise in a wide range of applications. Since little
intuition has been developed about how the results are affected by the
relevant experimental variables, e.g., cluster type, mass, and energy,
these attempts at establishing feasibility may be quite far from being
optimal in the choices of these parameters. I propose here that simulations based on molecular dynamics techniques may play a very important
role in the efficient design and accurate interpretation of the next
generation of such experiments. In addition to the impact this will have
on applications in materials science, these calculations and the relevant
experiments that follow may be important in determining the lifetimes of
small grains in the interstellar medium.

I. INTRODUCTION

Accelerated beams of large molecular ions (cluster ions) have been
used in a variety of feasibility experiments for applications that include
surface analysis [1], film deposition [2], micro-machining of surfaces
[3], and thermonuclear reactions [4]. Although it is still too early to
reach a final judgement concerning any unique contributions that cluster
ions will make to such applications, there have been enough positive
results already that significant further growth of this field is likely.
It is not my intention to provide here a complete review of the status of

---

*Supported in part by the National Aeronautics and Space Administrative
[NAGW-2279].

such investigations. The proceedings of a conference on this subject, which was held in Orsay in the summer of 1988, review the literature up to that point [5]; the examples given here represent a partial updating of that list.

One naively expects that when a number of energetic atoms are incident simultaneously on a small area of a surface (10-100 $\overset{\circ}{A}^2$) collective phenomena will arise that are not necessarily present in single ion impacts. The experimental results that have been obtained so far confirm such expectations: for example, a large enhancement in secondary ion yields [1]; high-quality film growth [2]; extremely smooth, highly polished surfaces [3]; and an apparent enormous enhancement of thermonuclear reaction yields [4]. In several of these cases, however, questions have arisen about whether these results should be attributed to other uncontrolled parameters in the experiments (e.g., the presence of lower mass, higher velocity ions contaminating the incident beam) rather than being due to the intended cluster ion irradiation. At this early stage in the development of this technique, such disagreements are to be expected - our intuition of how effects scale with cluster type, mass, and energy is not well developed, and the experiments attempted are, therefore, probably more primitive than they should be.

We find ourselves in the situation where basic theory is probably overwhelmed by the non-linearities of these many-body processes, and experiment has too many variables to control efficiently - especially in the absence of simple rules for deciding the relative importance of each of the variables. This is where simulation may provide a crude bridge over which we may pass to design a next generation of experiments that are less compromised by the subtle flaws that may well be leading us astray at the present time.

In this paper I outline how molecular dynamics (MD) simulations of cluster ion-induced processes may be able to serve as an aid to the orderly expansion/exploitation of this field. These simulations should be viewed as replacing neither basic theory nor detailed experiment; rather they are meant in the sense of a caricature of reality to explore the parameter space of experimental variables so as to give us more efficient starting points for optimizing the technique in particular situations.

II. MOLECULAR DYNAMICS SIMULATIONS

For transient phenomena like those occurring in ion impacts (e.g., sputtering and implantation) these calculations have a relatively standard embodiment. All the atoms involved are assumed to interact through potentials, and although these potentials have buried within them the quantum mechanics of the electrons and nuclei involved, the evolution of the atoms in the simulation occurs through the classical equations of motion of the system. This is, of course, exactly the spirit of the Born-Oppenheimer approximation as expressed in the Feynman-Hellmann theorem. For the energies that occur there, eV to keV per atom, this is quite a reasonable approximation [6].

If there are N atoms in the simulated system, the number of two-body interactions among them goes as N^2. This is reduced in practice by the fact that interatomic potentials decay rapidly with separation and can be neglected at distances beyond a few atomic diameters. Nevertheless, even with very efficient computing algorithms the system size is usually limited to values of N between 10^3 and 10^4. For cases where a particular moving atom interacts mainly with atoms at rest this limitation can be overcome by considering only the binary interactions of the moving atom - as in TRIM [7]. However, the simultaneous impact of several atoms in a small region of the target that occurs for a cluster ion makes such an approximation untenable. Thus, one is left with the most unrealistic of the characteristics of molecular dynamics - the small size of the system that tends to over-emphasize surface effects relative to a real system.

The interaction of the atoms in the simulation is through their collisions, which are mediated by the potentials choosen. If, for the moment, we ignore many-body effects, the form of a particular atom-atom potential is known fairly well. Obviously, the atoms do not interact at large distances, and the shape is known from the spectra of diatomic molecules at separations corresponding to the minimum in the potential energy. In a solid the elastic moduli give the shape near this minimum. At very small distances the potential is repulsive - that of a screened Coulomb field. Between that region and the attractive minimum the potential is also repulsive and arises from the fact that electrons in the two atoms are in virtually identical states, i.e., the Pauli repulsion. This region is the least well specified, and since it corresponds to interactions at 10-100 eV/atom, this uncertainty is of possible significance in the simulations we discuss here [8].

In the case of quasi-static configurations (e.g., the structure of small molecules or bulk solids) one can exploit the Born-Oppenheimer approximation to do the ab initio calculations of quantum chemistry to develop highly accurate forms for the interatomic potentials. However, in our case where atoms can move large distances relative to one another in the course of a single simulation, these ab initio methods are for the present beyond the capability of even super computers. One needs, therefore, a recipe to improve on the simple pair potentials discussed in the previous paragraph that allows for a smooth transition between the diatomic molecule-like interaction of a pair of atoms and its form when they are embedded in a solid. Obviously, this recipe should be able to deal with a region of low symmetry, for example, as exists for atoms near the surface.

It has been shown that the energy of an atom in a material is a function only of the local electron density - an effective medium approach which builds in the effects of many-body interactions [9]. A simplified version of this is given by the embedded atom method (EAM) [10], which takes spherically-averaged wave functions for the isolated atoms and adds their absolute squares together to estimate the local electron density in the many-body situation. The embedding function, i.e., the relation of the density to the interatomic potentials, is then constructed phenomenologically from universal binding considerations [11] and/or the detailed values of diatomic molecular spectra and elastic moduli [12]. Although the EAM recipe works in many situations, it leaves out the angle dependence of chemical bonds (e.g., it is spherically averaged and thus tends to do best with close-packed structures) and therefore does not always allow a particularly smooth transition between the diatomic molecular potential and the potential shape needed to give the correct elastic moduli of the solid [13].

One could from these remarks decide that the choice of interatomic potentials represents a major problem. This is certainly true if one wants to mimic reality exactly; if, however, we retreat to the original goal merely of developing a plausible cartoon of reality to guide intuition, then we find that the "observations" from simulations with different potential construction recipes are qualitatively similar.

III. EXAMPLES OF CLUSTER-ION SIMULATIONS

A. Dimer Impacts

It has beeen known for some time that the sputtering yield of a heavy diatomic molecular ion (dimer ion) is greater than twice that of a single ion impact with the same velocity. This is attributed to the fact that depositing so much energy in a small volume causes the linear relation of sputtering yield to deposited collisional energy to break down because atoms set in motion by one atom's impact interact with moving atoms from the other impact and not just with stationary atoms of the target.

In a series of simulations involving dimers, Mark Shapiro and I showed that although Ar_2^+ and Cu_2^+ produced no enhancement in sputtering yield, Kr_2^+ and Xe_2^+ gave yields 2.3 and 2.2 times as great, respectively, as single ions with the same velocity [14]. In Figure 1 is shown the sputtering yield obtained for 5 keV/atom single ion and dimer impacts on

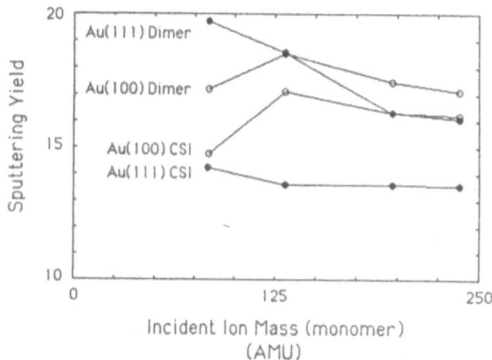

FIGURE 1: Sputtering yields for 5 keV/atom single ions (CSI) and dimers on (111) and (100) Au surfaces as a function of the mass of the incident atom. The calculations for the single impacts are the average of uncorrelated impacts at the same impact points as the two atoms in the dimer [15].

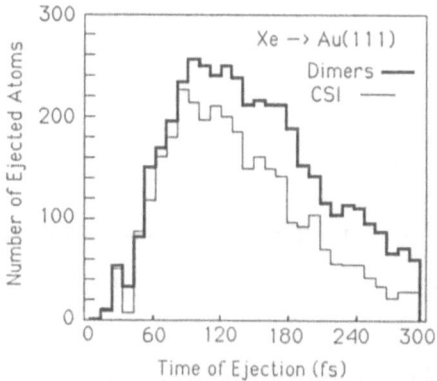

FIGURE 2: Ejection times for atoms sputtered by 5 keV/atom Xe dimers and combined single-ion impacts on Au(111). The calculations for the latter are the sum of uncorrelated impacts at the same impact points as the two atoms in the dimer [15].

(111) and (100) Au surfaces. Although these results indicate clearly the presence of non-linearities due to the dimers, the increase in the single-ion yield at higher ion masses on the more open (100) surface also shows a growing non-linear contribution ("ion spike") to the sputtering [15].

In Figure 2 we show the microscopic nature of this enhancement [15]. Plotted here is the number of sputtered atoms versus time of ejection; although the numbers are virtually identical for the first 90 fs, more atoms escape at later times for the dimer impact. Thus, one could say crudely that the spot of impact stays "hot" longer for a dimer impact than for two uncorrelated impacts at the same impact points. These simulations also show that the energy spectrum within the cascade of moving atoms extends to higher energies - as one expects if moving atoms collide with other moving atoms and not just stationary ones.

B. Simulated Impacts at 1 keV/atom

The various attempts to induce thermonuclear-reactions by the impact of cluster ions containing deuterium atoms have provoked considerable controversy [4,16,17]. Whatever the ultimate outcome of these investigations, it is of interest to know what happens when cluster ions in this velocity range strike surfaces. At the very least this is the same sort of velocity (~ 100 km/sec) as that produced when a supernova-induced shock wave passes through a molecular cloud in the interstellar medium [18]. The survival probability in grain-grain collisions in such an environment is not understood well and has a direct bearing on the interpretation of some meteoritic observations [19] and the role such grains may play as the sites for chemical reactions in the interstellar medium that produce some types of organic molecules [20].

In Figure 3 are shown the local density versus time obtained by Shapiro and myself for simulated impacts of Al_{32} and Al_{63} clusters on Al and Au targets [21,22]. Although the density reaches about twice the original value, the system is unconfined and comes apart in ~ 20 fs. The kinetic energy spectrum (i.e., temperature) during this period is also determined in the calculation, and we find ourselves very far from a situation where there would be a measurable enhancement in the thermo-nuclear reaction yield for appropriate atomic species.

Figure 4 shows what we think is the origin of the nuclear reactions detected experimentally [4]. Here is shown the energy spectrum of atoms in the projectile that rebound from the target surface - a "hot splash" of Al in our simulation. For pure Al clusters $\sim 25\%$ of the atoms rebound;

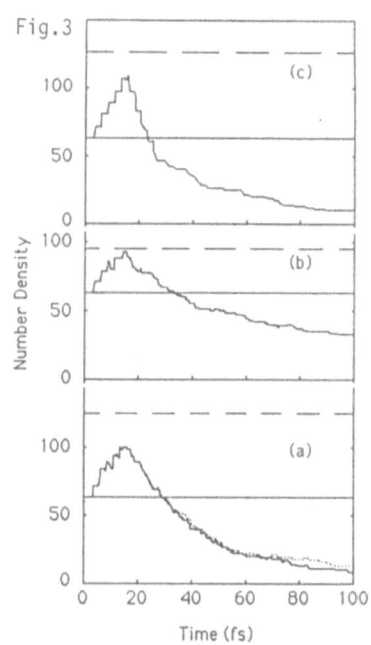

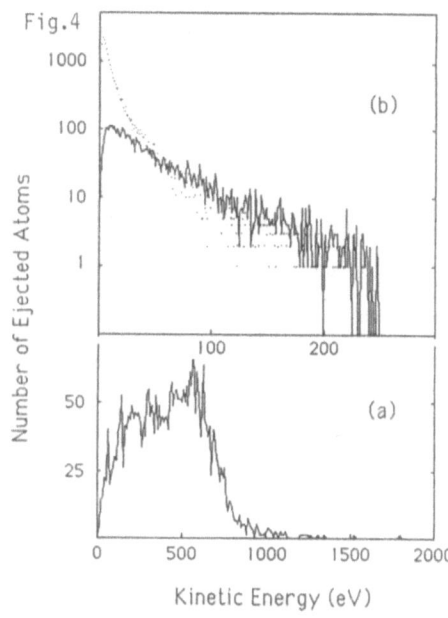

FIGURE 3: Number density vs time plots for typical cluster impacts at 1 keV/atom for the (a) Al_{63} on Au, (b) Al_{32} on Au, and (c) Al_{63} on Al systems. The dashed curve in (a) was computed with a 12-layer target. The solid horizontal lines represent the initial number density of the target, and the horizontal dashed lines represent the number density that would be reached if all cluster atoms were confined with no loss of target atoms. The vertical scales in (a) and (b) should be multiplied by 9.38×10^{26} m^{-3}, and that in (c) should be multiplied by 9.69×10^{26} m^{-3} to obtain absolute number densities [21].

FIGURE 4: Energy spectra for (a) aluminum and (b) gold atoms ejected during the first 100 fs in Al_{63} impacts on gold targets (solid lines). The energy spectrum for gold atoms ejected during the first 500 fs is shown dotted. Note the different energy scales [21].

for a mixed cluster ($Al_{38}Au_{25}$) $\sim$ 50% of the Al atoms rebound. For the

$(D_2O)_n^+$ beams used by Beuhler, et al. this component would be thrown back

(mainly as individual atoms) into the ion accelerator, since there is a

line-of-sight path between target and accelerator in this experiment [4].

Some of these deuterons would be ionized and reaccelerated - causing a low

mass, high velocity contamination in the beam that would preferentially

produce the observed nuclear reactions (e.g, d + d → p + t). In the

experiment of Fallavier, et al. with D_n^+ clusters in which no nuclear reactions were observed, there was mass analysis between target and accelerator, and thus there was no opportunity for this process to occur [16].

In all our simulations, there is a considerable enhancement of the sputtering yield of target atoms over that produced by uncorrelated impacts of the same number of single ions of the same velocity. Because the simulated systems are small (2000-4000 atoms) the energy from the impact is not contained completely in the target; thus, the yields for the heavy clusters are underestimated to an increasing degree with increasing cluster size. Even with this in mind there does seem to be an eventual saturation of the enhancement with increasing cluster size. This was described in the MD simulation work of Shulga and Sigmund as "clearing the way", i.e., the first atoms to impact remove target atoms from the path of subsequent atoms that impact from the same cluster [23].

C. Low-energy (eV/atom) Cluster Ion Impacts

It is important to improve our intuition about low-energy impacts because of the possible application of the technique to epitaxial film growth at low temperatures [1]. The work of Hsieh and Averback in simulating Cu cluster impacts on Cu (100) provides an important beginning to this work [24]. They found that small clusters with energies of 25 eV/atom create small craters, whereas larger clusters at $\sim$ 3.5 eV/atom form epitaxial layers on the surface without creating point defects. In neither case do atoms from the cluster seem to break away and migrate long distances over the substrate surface. Our own work for 1-10 eV/atom Al_{32} and Al_{63} clusters on Au yields results that are very similar [25]. We thus see that the simulations provide a guide both to interpretating the observations and suggesting the relevant cluster size and energy that should be used.

IV. FUTURE DIRECTIONS

It seems clear already that we can answer qualitatively some of the
questions raised about interstellar grain collisions. At 100 km/sec ($\sim$ 1
keV/atom) there is enough rebound and sputtering to disrupt completely the
grains involved; however, at 10 km/sec ($\sim$ 10 eV/atom) the grains stick
together with a minimal chance of losing atoms. The velocity range between is particularly interesting because it corresponds to the sort of
shock wave environment to which a typical grain may well have been exposed
in the time between its birth in the atmosphere of an asymptotic giant
branch star and its eventual incorporation into a forming solar system.
Carrying out MD simulations in this regime and comparing them with suitable laboratory experiments is, therefore, likely to define more sharply
the survival probability of these grains and the role they play in preser-
ving the signature of their stellar creation.

In studying cluster-ion impacts one should certainly plan to measure
differential sputtering yields for target and projectile atoms, but what
we need most is a diagnostic technique that tells more directly what is
really happening during the impact. In the case of a single incident ion,
by following each atom in the simulation and recording its distribution of
collision impact parameters, Shapiro and Fine were able to estimate semiclassically its probability of electronic excitation [26]. For small
impact parameters this leads to excited states that decay by x-ray and
energetic Auger electron emission that are easily detected. In Figure 5
we show the potential energy distributions for a number of cluster ion-
induced collision cascades [21,22]. Note that the two-body energy limit
for an incident Al in the cluster striking a stationary Au atom is 241 eV;
thus, multiple collision processes are able to raise this significantly.
There are a sufficient number of additional hard collisions in the cascade
that, although there is no prediction of appreciably enhanced **nuclear**
reactions, there is likely to be a significant increase in the number of
atomic excitations. By observing the x-ray or Auger electron yield versus

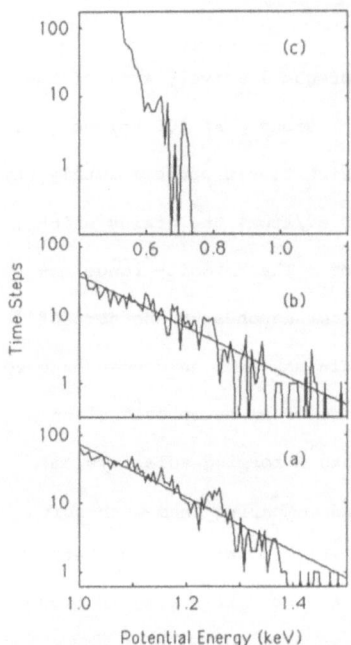

FIGURE 5: Number of time steps at which given potential energies per atom were reached for the (a) Al_{63} on Au, (b) Al_{32} on Au, and (c) Al_{63} on Al systems. Only results for aluminum atoms are shown in (a) and (b). The vertical scale in (a) should be divided by 200 to obtain time steps per impact. Similarly, those in (b) and (c) should be divided by 300. Note the different energy scale in (c). The solid lines represent least squares fits of decaying exponentials to the simulation results [21].

cluster size and energy an indirect measure of local atomic density and temperature can thereby be obtained. Shapiro and I have just begun to sort our exisiting simulation runs to provide more quantitative estimates of these excitation probabilities. Since the collisions that lead to those energies above the two-body limit only occur early in the cascade, one would by such observations probe the high-density, short-lived part of the impact process.

Another important ingredient in the expansion of cluster-ion applications is the design of efficient ion sources, accelerators, and mass analysis systems. One is going to need intense beams that are tightly controlled in the purity of ion type, mass, and energy if the dis-

54

crepancies that now exist are to be removed. MD simulations can be an important factor in deciding on realistic specifications for such accelerator developments.

V. ACKNOWLEDGEMENTS

The author is grateful to his collaborators, Jon Pelletier and Mark Shapiro, for access to their recent pioneering work in this field. Without their dedication to the elimination of spurious computational errors and to the improvement of computing efficiency, calculations of this sort could never have been performed.

I also am most favored to have been chosen as a participant in this dual birthday celebration. Not only have I been a collaborator for many years with members of Bethge's ion beam applications group, but at an earlier stage of my career that involved accelerator design I was also much influenced by Klein's outstanding work on helical linacs. It is, therefore, a double pleasure for me to be a contributor to this symposium. One can certainly see in the programs of these men the very ingredients we shall need to make further progress in the field I have described.

REFERENCES

1. M. G. Blain, S. Della-Negra, H. Joret, Y. LeBeyec, and E. A. Schweikert, Phys. Rev. Lett. 63, 1625 (1989).

2. I. Yamada, H. Taksoka, H. Usui, and T. Takagi, J. Vac. Sci. Techol. A4, 722 (1986).

3. P. R. W. Henkes and R. Klingelhofer, J. de Physique C2, 159 (1989).

4. R. J. Beuhler, G. Friedlander, and L. Friedman, Phys. Rev. Lett. 63, 1292 (1989).

5. Second International Workshop on MeV and keV Ions and Cluster Interactions with Surface and Materials, ed. Y. Le Beyec, S. Della-Negra, and J. P. Thomas, J. de Physique 50 (1989).

6. D. E. Harrison, Jr., Crit. Rev. Solid State and Matter Sci. 14, S1 (1988).

7. J. F. Ziegler, J. P. Biersack, and U. Littmark, The Stopping and Range of Ions in Solids, Vol. 1, Chapter 4 (Pergamon Press, 1985).

8.   P. J. Van Den Hoek, A. W. Kleyn, and E. J. Baerends, Comments At. Mol. Phys. **23**, 93 (1989).

9.   H. Hohenberg and W. Kohn, Phys. Rev. **136**, B864 (1964).

10. M. S. Daw and M. I. Baskes, Phys. Rev. Lett. **50**, 1285 (1983).

11. J. H. Rose, J. R. Smith, F. Guinea, and J. Ferrante, Phys. Rev. **B29**, 2963 (1984).

12. S. M. Foiles, M. I. Baskes, and M. S. Daw, Phys. Rev. **B33**, 7983 (1986).

13. B. J. Garrison, N. Winograd, D. M. Deaven, C. T. Reimann, D. Y. Lo, T. A. Tombrello, D. E. Harrison, Jr., and M. H. Shapiro, Phys. Rev. **B37**, 7197 (1988).

14. M. H. Shapiro and T. A. Tombrello, Nucl. Instr. Meth. **B48**, 557 (1990).

15. M. H. Shapiro and T. A. Tombrello, Nucl. Instr. Meth. **B** (1991) submitted.

16. M. A. Fallavier, J. Kemmler, P. Krisch, J. C. Poizat, J. Remillieux, and J. P. Thomas, Phys. Rev. Lett. **65**, 621 (1990).

17. C. Carraro, B. Q. Chen, S. Schramm, and S. E. Koonin, Phys. Rev. **A42**, 1379 (1990).

18. B. T. Drain and E. E. Salpeter, Ap. J. **231**, 438 (1979).

19. R. S. Lewis, S. Amari, E. Anders, Nature **348**, 293 (1990).

20. W. D. Watson, Rev. Mod. Phys. **48**, 513 (1976).

21. M. H. Shapiro and T. A. Tombrello, Phys. Rev. Lett. **65**, 92 (1990).

22. M. H. Shapiro and T. A. Tombrello, Nucl. Inst. Meth. **B** (1991) in press.

23. V. I. Shulga and P. Sigmund, Nucl. Inst. Meth. **B47**, 236 (1990).

24. H. Hsieh and R. S. Averback, Phys. Rev. **B42**, 5365 (1990).

25. J. Pelletier, M. H. Shaprio, and T. A. Tombrello, Nucl. Instr. Meth. **B** (1991) submitted.

26. M. H. Shaprio and J. Fine, Nucl. Instr. Meth. **B44**, 43 (1989).

Channeling: A Marriage of Atomic Collision Physics and Materials Science

S. Datz

Manne Siegbahn Institute of Physics, Stockholm, Sweden, and
Oak Ridge National Laboratory, Oak Ridge, USA

Ever since the modern "discoveries" of channeling in *ca* 1962 [1−4] there has been a close connection between atomic collision physics and the science of materials. Channeling, the phenomenon which leads to steering of a penetrating ion's trajectory along low index crystal directions, is caused by correlated atomic collisions with the atoms lying in crystal rows or planes. Without crystalographic order there can be no channeling, hence a method for measuring disorder, defects, interstitials etc.

On the other hand channeled ions are steered away from close collisions with individual lattice atoms and results in a reduction in nuclear reactions, Rutherford scattering, inner-shell ionization and stopping power. This steering effect also causes the penetrating ion to collide only with loosely bound conduction or valence electrons. When the ion's velocity greatly exceeds the Fermi velocity of the conduction electrons, the crystal channel may be used as an atomic collisions laboratory in which enormous electron fluxes are obtained similar to those encountered in dense plasmas but with a much smaller momentum distribution and a specific directionality.

To review all of the aspects of this subject is well beyond the scope of this article. Instead I will take a few examples sampling both material science and atomic collision physics.

Surface Melting

Recent work at the FOM Institute for Atomic and Molecular Physics in Amsterdam used the technique of medium-energy ion scattering (MEIS) to measure surface melting below the melting point of the bulk material[5,6]. This has long been suspected but direct proof has been lacking. In the MEIS technique ~ 100 keV protons are backscattered from the crystal under investigation. The crystal is alligned in a low index direction so that if crystalographic order is maintained all ions will penetrate the crystal (channel) and stop; except those which strike a surface atom. These will be reflected and their final energy is determined by the depth in the solid from which they have been scattered.

This is illustrated in Fig. 1. The results for backscattering from an Al(110) surface are shown in Fig. 2. As the temperature is increased the order and thus the backscattering yield and energy distribution increases due to increased lattice vibration and finally at the melting point all order disappears. This is shown in Fig. 3. The solid line sows the anticipated behaviour simply due to increased vibrational amplitude. The deviations starting at *ca* 800° indicate surface melting. In previous studies by the same group it was shown that the surface melting in Pb was dependent upon crystal face. Open faces such as Pb(110) exhibiting a surface melting transition while the more close packed (111) face remained ordered all the way up to the bulk melting point.

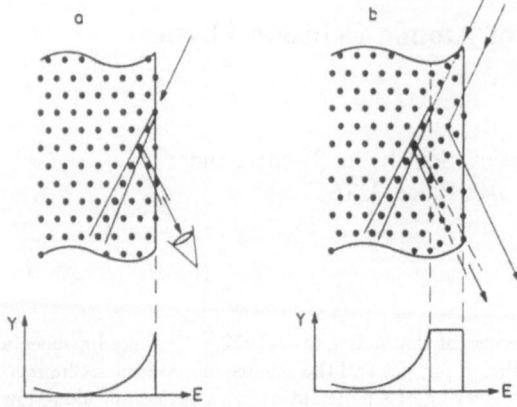

Figure 1

Schematic representation of energy spectra for (a) a well-ordered crystal surface, and (b) a crystal covered by a disordered surface layer. Ion beam and detector are aligned with the $[\bar{1}01]$ and $[011]$ directions in the $(1\bar{1}1)$ crystal plane. Shadowing and blocking effects are indicated.

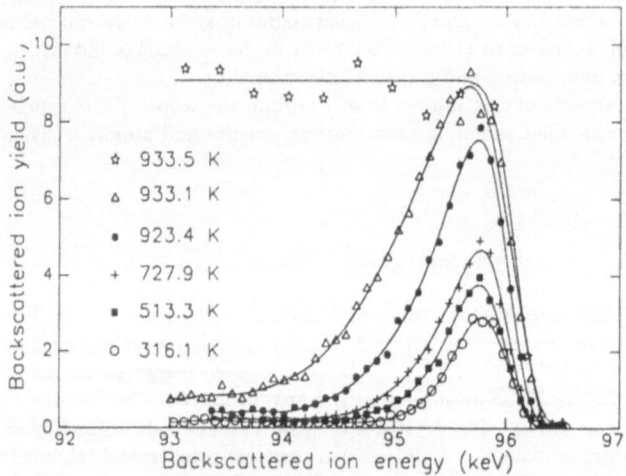

Figure 2

Backscattering energy spectra from the Al(110) surface taken at various temperatures in the range from 316 to 933.5 K.

58

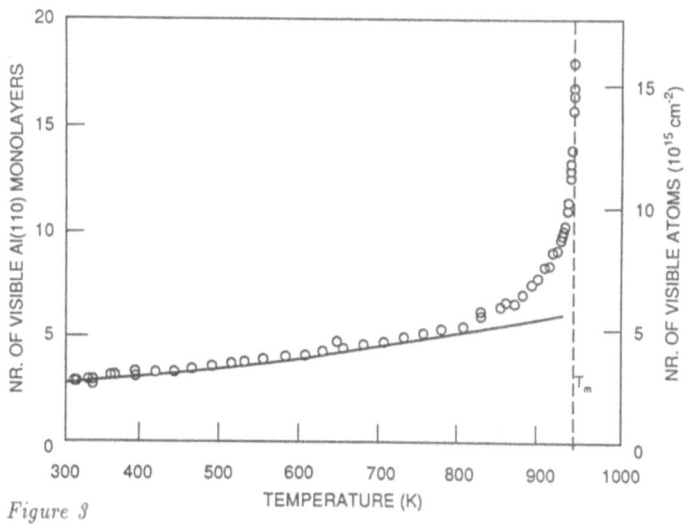

Figure 3

Surface peak area as a function of temperature for the Al(110) surface. The solid curve shows the yield expected for an ordered Al(110) surface, as calculated in a Monte Carlo simulation of the experiment. The deviation of the experimental data from the simulated yield is due to surface melting.

Coulomb Probes of Crystal Potentials

The motions of ions in channels are determined by the crystal potential, the potential is directly related to the local electron density. Since the stopping power for the ion is determined by the electron density, measurements of the inter-relation between ion trajectory and stopping power can yield mappings of the crystal potential[7,8].

Since most particles moving in planar channels of a crystal do not enter the crystal along the midplane, they generally oscillate transversely about the midplane between two sheets of atoms. Correlated small-angle scatterings cause the ions to return to (and cross) the midplane, executing transverse oscillations as shown in Fig. 4. If the beam of particles and the detector are aligned and well collimated, the detector will pick out groups of the trajectories which return the particles to the beam axis. Particles which do not have an integral number of oscillations will not be detected because their emergence angles will not coincide with that of the detector. For very small oscillation amplitudes, the particles will be able to reach the detector even if they have a nonintegral number of transverse oscillations because all emergence angles for those amplitudes are within the detector acceptance angle. Near the center of the channel the interplanar potential is such that the oscillation frequencies become constant. In the harmonic region the frequency remains constant but the stopping power continues to fall very gradually to the minimum value which occurs at zero oscillation amplitude. Particles with larger amplitudes will have larger stopping powers since they travel in regions of greater electron density. The wavelength becomes shorter for larger amplitude so that wavelength, oscillation amplitude, and average stopping power are related. They are governed by the details of the restoring field. These effects cause the energy-loss spectrum to have series of peaks, each peak corresponding to an integral number of wavelengths. If the crystal is tipped slightly, each trajectory will start at a different

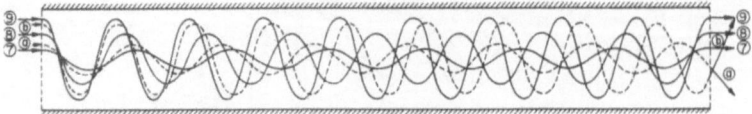

CHANNEL "WALL"

1. Detector Aligned to Beam Axis.
2. Paths ⑦ ⑧ and ⑨ Emerge with Direction Unchanged.
3. Paths Such As ⓐ and ⓑ Emerge with Altered Direction.
4. An Identical Set of Paths (Omitted Here) Exists for Entry Below Mid-Channel.
5. $(\Delta E)_9 > (\Delta E)_8 > (\Delta E)_7$ etc.
   $(\Delta E)_a$ and $(\Delta E)_b$ Are Unobserved.
6. For the Case Drawn Here λ Decreases with Oscillation Amplitude.

Figure 4

Selection of energy loss groups according to angle of emergence of channeled particles.

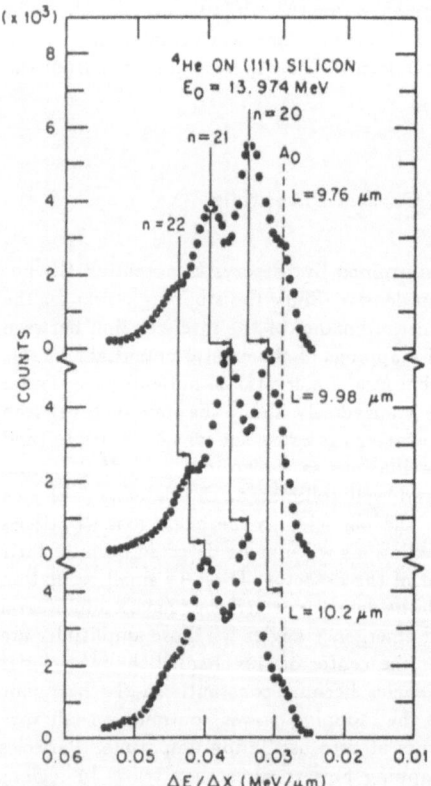

Figure 5

Shifts of planar channel energy-loss groups vs crystal thickness.

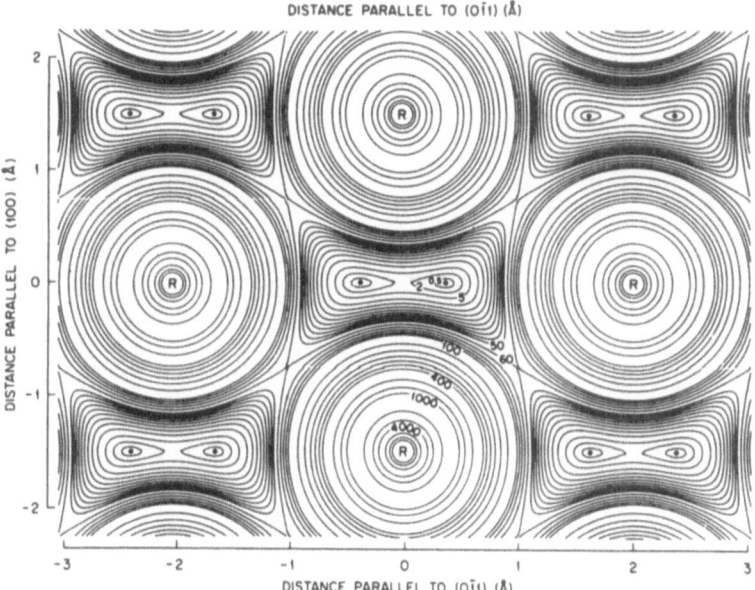

Figure 6

Potential energy countours of I ions in the continuum potentials of ⟨011⟩ rows (R) of Ag atoms. All values are in eV and are relative to the minimum (•), which has an absolute value of 72 eV.

point on the wave and very small oscillation amplitudes will be eliminated. Further tipping eliminates more small amplitudes of oscillation and will accentuate the larger-amplitude, short-wavelength, large-stopping-power, groups. If the crystal is rotated about an axis perpendicular to the planes, the path length in the crystal will be increased and the geometry will select a series of longer wavelengths with smaller stopping powers. Crystal rotation can give enough path-length variation to bring a given group with one oscillation number to the same stopping power as had been obtained previously for the adjacent group (see Fig. 5). Thus one can map out the stopping power versus wavelength continuously over a wide range of oscillation amplitudes.

In the particular study referred to in Fig. 5 ions of ^{3}He, ^{4}He and ^{6}Li and ^{7}Li at 3.5 MeV/u were used to map planar channels in Si. The particular choice of ions was made to show the independence of the measured potential of Coulomb probe charge, mass and velocity.

Similar measurements[9], too lengthy to describe here, in which trajectories in axial channeling were used have led to potential maps such as the one shown in Fig. 6 for the ⟨110⟩ direction in Ag.

"Frozen" Charge States

Ion beams passing through solid targets undergo collisions which lead to both capture and loss processes. Under normal conditions these processes are so rapid that the charge state distribution of the emergent ions is independent of the input charge. This is illustrated in Fig. 7 in the distribution for 315 MeV Ti ions emergent from a 2.6 μm Si crystal in a random orientation, i.e. we achieve charge state equilibrium. When the

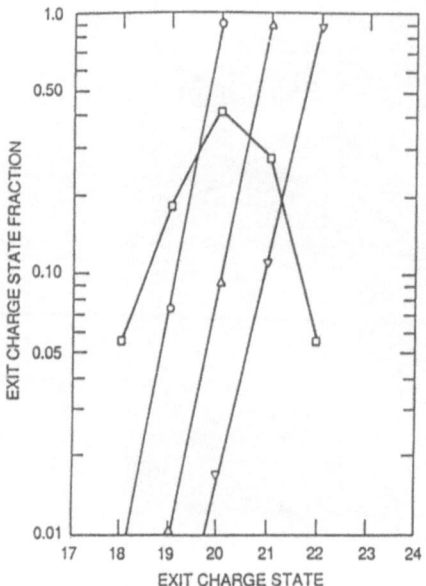

Figure 7

Exit charge state distributions for injection of 315 MeV Ti^{21+} (o), Ti^{21+} ($\triangle$), Ti^{22+} ($\triangledown$) in a $\langle 100 \rangle$ channel in Si 2.6μm thick, and ($\square$) exit charge distribution for random direction.

crystal is oriented to the $\langle 100 \rangle$ axial direction the situation is quite different. The exit charge state distribution is strongly dependent upon input charge[10,11]. Charge capture is limited to $\sim 10\%$ and charge loss is too small to indicate on this scale. This comes about beause of constriction to interact only with valence electrons. The ion velocity ~ 7 MeV/u corresponds to an electron velocity of ~ 3.5 keV and the probability of capture of such semi-free electrons is small. On the other hand the ionization potential of e.g. Ti^{21+} is ~ 6.6 keV so that electron bombardment energy is much below threshold. Thus we have taken a "thick" target for charge transfer and converted it into a "thin" target from which we may derive single collision cross sections.

Radiative Electron Capture (REC)

In order to determine cross sections we must have a knowledge of the electron density and the crystal thickness. One way of measuring the density (a characteristic of the material) is by measuring the width of the radiative electron capture peak (an atomic collision process). In radiative electron capture (REC) an electron from the continuum drops into a bound state and is stabilized by the emission of a photon. The photon energy for radiative capture to a 1s state is obtained from the sum of the 1s binding energy plus the kinetic energy of the electron in the continuum. Since the electrons in the channel have a Fermi energy distribution this will be reflected in the width of the REC peak[12]. Furthermore, for a Fermi gas, the energy distribution is determined by the electron density. Hence a direct measure of the density and hence region of the crystal involved. An example of this is shown in Fig. 8. Here we have measured X-ray spectra in coincidence with electron capture for 315 MeV Ti ions passing through the

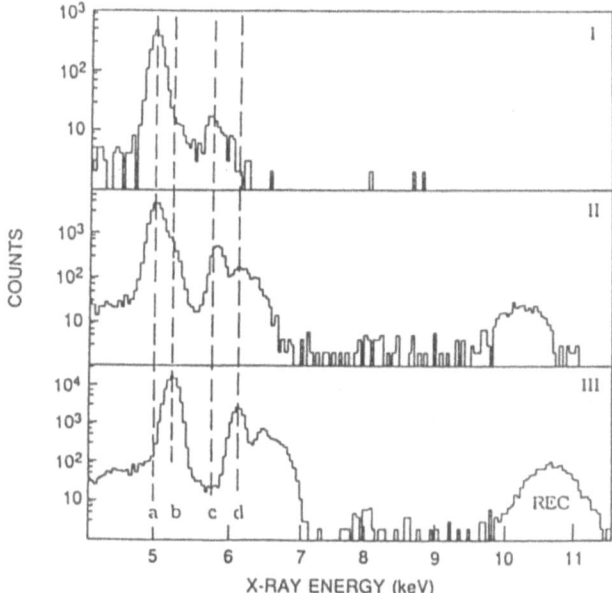

Figure 8

X-ray spectra (singles) for 315 MeV Ti^{q+} channeled through $\langle 100 \rangle$ Si 2.6μm thick: (I) $q = 20$, (II) $q = 21$, and (III) $q = 22$ taken in coincidence with emergent charge state $(q-1)$. Specific features identified in the spectra are, respectively: (a) $1s2\ell \rightarrow 1s^2$, (b) $2p \rightarrow 1s$, (c) $1s3\ell \rightarrow 1s^2$, and (d) $ep \rightarrow 1s$. K-radiative electron capture X-rays appear at 10.3 keV in II and 10.7 keV in III.

$\langle 100 \rangle$ channel of a 2.6 μm thick Si crystal[11]. The bottom part of the figure bare ions are injected i.e. Ti^{22+}+ $\rightarrow$ Ti^{21+}(1s)+$h\nu$. The REC peak appears in the 10 keV region. In the middle portion of the figure using hydrogenic ions i.e. Ti^{21+}(1s)+e $\rightarrow$ Ti^{20+}(1s)+$h\nu$ the REC peak is shifted down because of the screening by the 1s electron. Finally using Ti^{20+} injection no K-REC peak is seen because no K hole exists. From the shape and width of the REC peaks we obtain a Fermi width of 10 eV in accord with theory.

Dielectronic Excitation in Channels

Here we make use of the channeling phenomenon to study interactions of ions in a dense electron gas. In particular dielectronic excitation[11]. Take for example a hydorgenic ion colliding with an electron to form a dielectronically excited state $2\ell2\ell'$

$$A^{(z-1)+}(1s) + e \rightarrow [A^{(z-2)}(2\ell2\ell')]^{**} \tag{1}$$

The reverse of this reaction is a KLL Auger event. In a dilute medium we will either get Auger relaxation from this state (reverse of eq. 1) or radiative stabilization leading to recombination

$$[A^{(z-2)+}(2\ell2\ell')]^{**} \rightarrow [A^{(z-2)+}(1s2\ell')]^* + h\nu_1 \rightarrow A^{(z-2)+}(1s^2) + h\nu_2 \tag{2}$$

In a very dense gas, however, (the electron density in the channel is $\sim 10^{23}$ cm^{-3})

63

second and even third electronic collisions can occur before the dielectronically excited ion either radiates or escapes from the crystal. For example collisional ℓ mixing is very rapid.

$$[A^{(z-2)+}(2\ell 2\ell')]^{**} + e \rightarrow [A^{(z-2)+}(2\ell' 2\ell'')]^{**} + e \qquad (3)$$

Collisional excitation or ionization

$$\text{e.g.}\,[A^{(z-2)+}(2\ell 2\ell')]^{**} + e \rightarrow [A^{(z-2)+}(2\ell 3\ell)]^{**} + e \qquad (4)$$

$$[A^{(z-2)+}(2\ell 3\ell)]^{**} + e \rightarrow [A^{(z-1)}2\ell]^{*} + 2e \qquad (5)$$

this can be followed by radiative decay

$$[A^{(z-1)+}(2\ell)]^{*} \rightarrow A^{(z-1)+}(1s) + h\nu_3 \qquad (6)$$

or by yet another electronic collision which, in fact, gives a net result of ionization rather than recombination

$$[A^{(z-1)+}(2\ell)]^{*} + e \rightarrow A^{z+} + e \qquad (7)$$

(Note that the resonant energy for the formation of the dielectronically excited state (eq. 1) is $\sim 1/2$ IP and lies below even the direct excitation threshold (3/4 IP)).

Magnitudes of these contributions can be estimated by assuming rapid collisional substate mixing, $n\ell \leftrightarrow n\ell'$ and then using the fastest available radiative and Auger rates. The radiative and Auger rates have been tabulated by Seely[13]. The ionization cross sections are estimated from the Lotz semi-empirical formula[14]. The excitation cross sections used are obtained from the Seaton formula.

No calculations exist for the cross sections for fine structure mixing (eq. 3) for the ions in question. However, if we use the recent calculation[15] on, e.g., $(2s^2) \rightarrow (2s2p\,^3P_2)$ transition in Fe^{22+} we obtain a cross section of $\sim 3 \times 10^{-16}$ cm^2 at a velocity corresponding to KLL excitation of Fe^{23+}. Since the fine structure splitting decreases with decreasing Z, the corresponding cross sections for the ions in question should be even higher. The main point is that these cross sections are so large that complete mixing is assured. We have no direct calculation for the dielectronic excitation cross section (i.e., eqn. 1) instead, we estimate it from measurements of RTE cross sections on the same or similar systems. The collisional rates, r_i, are obtained from

$$r_i = \sigma_i\,\rho_e\,v_i$$

where ρ_e is the electron density taken to be 1.44×10^{23} cm^{-3} for the Si $\langle 110 \rangle$ channel and σ_i is the cross section for the considered process. However, since the dielectronic excitation process (eqn. 1) is a resonant one, the actual rate is grossly reduced (factor of ~ 1000) due to the Compton spread in relative energies. In our calculation, we use a 10 eV Fermi distribution for the electrons in a $\langle 110 \rangle$ Si channel. This value is taken from the work of Appleton et al.[12], and we derived a similar Fermi width from our measurement of the radiative electron capture (REC) peak.

In fig. 9a, we show for channeled Ca^{19+} the yield per injected ion of $(h\nu_1 + h\nu_3)$ as a function of ion energy. The features correspond to dielectronic excitation of KLL, KLM, KLN, etc., and, at 300 MeV, to direct $1s \rightarrow 2p$ excitation. The energies at which these contributions should appear are indicated by the arrows at the bottom of fig. 9a. The solid curves in fig. 9a are calculated from rate equations shown in eqns 1–9, folded with the Compton profile of the 10 eV Fermi distribution. The populations of the ten lowest hydrogen- and helium-like configurations ($1s$ to 3ℓ and $1s^2$ to $3\ell 3\ell'$, respectively) are iteratively solved and summation bins for numbers of X-rays emitted are accumulated in 400 steps through the crystal length of 1.2 μm. charge-state fractions and x-ray

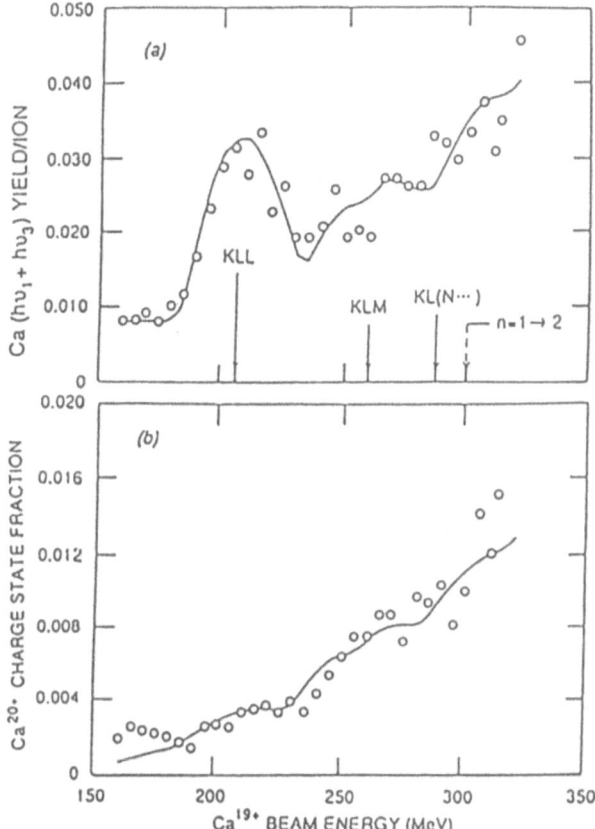

Figure 9

(a) Yield of $h\nu_1(2\ell n\ell' \rightarrow 1sn\ell')$ plus $h\nu_3(2\ell \rightarrow 1s)$ calcium X-rays and (b) charge fraction of Ca^{20+} ions as a function of Ca^{19+} ion energy incident on a $\langle 100 \rangle$ channel in Si (1.2 μm thick). The smooth curves are results of a computer simulation as described in the text.

yields per ion are calculated, including effects of Auger and X-ray decay of excited ions downstream of the crystal. It should be noted that the only normalization is the addition of backgrounds linear in projectile energy and that the predicted values are absolute in X-rays per ion. Using the same rate coefficients, we can calculate, e.g., the yield of Ca^{20+} ions (cf. eqn. 1-V.b) and plot it together with the experimental data in fig. 9b.

In the case of Ca^{19+}, the radiative processes dominate, e.g., the ratio of rates for the first branch point $\simeq 5$ (i.e. (eq. 2) vs (eq. 4)). In the case of S^{15+}, the radiative rates are slower and the excitation and ionization cross sections are higher so that the collisional channel comes more into play, i.e., for $S^{15+} \simeq 1$. In fig. 10a, we show, as a function of S^{15+} ion energy, the data for the yields of $(h\nu_1 + h\nu_3)$. In fig. 10b we show the S^{16+} charge fraction which is correspondingly higher than the totally stripped ion fraction for Ca. The solid curves are calculated in the same manner as for Ca^{19+} above.

65

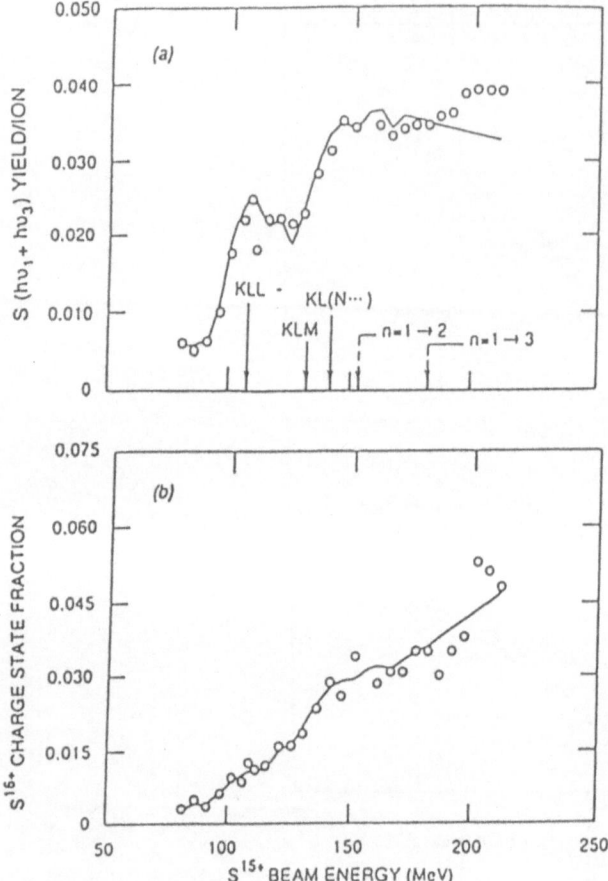

Figure 10

(a)Yield of $h\nu_1(2\ell n\ell' \rightarrow 1sn\ell')$ plus $h\nu_3(2\ell \rightarrow 1s)$ sulphur X-rays and (b) charge fraction of S^{16+} ions as a function of S^{15+} ion energy incident on a $\langle 100 \rangle$ channel in Si (1.2 μm thick). The smooth curves are results of a computer simulation as described in the text.

In the case of Ti^{21+} the ratio of rates at the first branch point $\cong 9$. Accordingly, the number of totally stripped ions becomes vanishingly small, but the effect is easily seen in the X-ray channel alone as demonstrated in fig. 11.

In all the above we have made many assumptions concerning rates. To validate these assumptions we take a simpler case where radiative processes should dominate (i.e. where secondary electron collisions are unimportant) and see whether we can derive a cross section for dielectronic recombination. For this we chose the case of the KLL resonance in Ti^{21+} and measured the X-ray spectra in coincidence with Ti^{20+}. This measures the dielectronic recombination rate. The importance of secondary collisions can be measured by varying the thickness of the Si crystal used i.e. in the absence of secondary collisions the yield will be linearly proportional to thickness. The results for the yield per unit thickness is shown for the three thicknesses used 1.1, 2.1, and 2.8 μm in fig. 12 and are sensibly independent of thickness. The curve in fig. 12 is for a

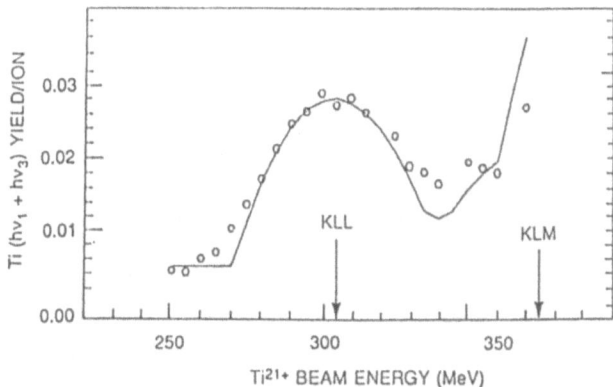

Figure 11

Yield of $h\nu_1(2\ell n\ell' \to 1sn\ell)$ plus $h\nu_3(2\ell \to 1s)$ titanium X-rays as a function of Ti^{21^+} ion energy incident on a $\langle 110 \rangle$ channel in Si (1.2 μm thick). The smooth curve is a result of a computer simulation as described in the text.

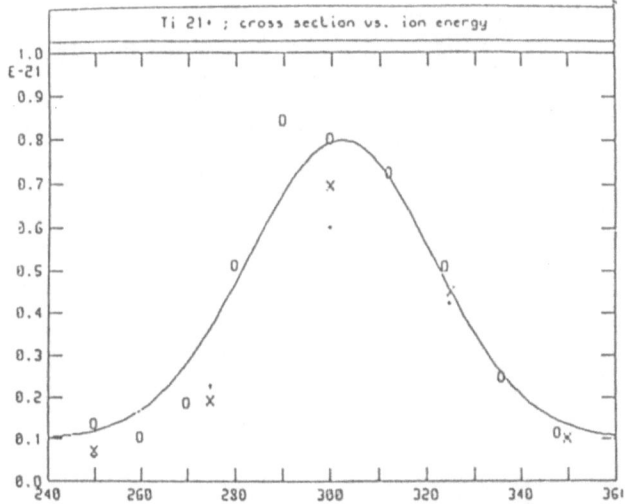

Figure 12

Coincidence measurements for $Ti^{21^+} + e \to Ti^{21^+} + h\nu$ in $\langle 110 \rangle$ Si as a function of Ti ion energy. Yield in the KLL resonance region as a function of crystal thickness; 0–1.1 μm, +2.1 μm, X 2.8 μm.

10 eV Fermi distribution fit to the points. From this we can derive a cross section of $0.70 \pm 0.10 \times 10^{-21}$ cm^2 per electron. To compare with RTE data (where the width of the resonance is determined by the Compton profile of H_2) we multiply our values by 1.2 and, since the RTE cross sections are given per molecule, we divide those values by two. The value thus obtained from the RTE data is 0.72 ± 0.12 i.e. there is excellent agreement between the channeling method and the RTE gas target measurements.

67

Resonant Coherent Excitation (RCE)

As a final note on the marriage of material science with atomic collision physics I mention some of our recent work on resonant coherent excitation. Here we use the periodicity of the crystal lattice to excite the channeled ion to a specific short lived state and then study the collisions of such ions with electrons in the channel.

Hydrogenic ions passing through axial and planar channels can be excited from $n = 1$ to $n = 2$ when the frequency of perturbation by the atoms in the crystal spaced a distance d apart came into resonance when the spacing between eigenstates i and j,

$$\Delta E_{ij} = hK(v_i/d)$$

where K is a harmonic $1,2,3 \ldots$ of the (v_i/d) frequency(16, 17, 18). In a crystal channel the degeneracy in the $n = 2$ levels is removed; first by the assymetry in the crystal field and second by Stark mixing of $2s$ with $2p_z$ (z being the direction of ionic motion) which is caused by the wake field following the ion. The main point in the context of this paper is that the resonant frequency, and hence velocity, for excitation to $2p_{x,y}$ is different than that for $2p_z$ and they can be excited selectively.

Once the ion is excited it is much more susceptible to collisional ionization by electron impact. Therefore the effect may be observed either by an enhancement in the ionized (totally bare) ions exciting the crystal or by an increase in the yield of Ly_α X-rays emitted from the $n = 2$ excited states of ions which have excited the crystal.

For the experiment presented here we used 4th harmonic excitation of Mg^{11+} [19].

The HHIRF at ORNL supplied beams of Mg^{11+} at energies from 145 to 170 MeV. These were passed through an Au crystal 1500 Å thick in the $\langle 111 \rangle$ directions (d = 7.06 Å). The emergent ion charge distribution and the Mg K X-ray yields were measured as a function of ion energy. The results of the X-ray measurements are shown in Fig. 13

RESONANT COHERENT EXCITATION OF Mg^{11+}

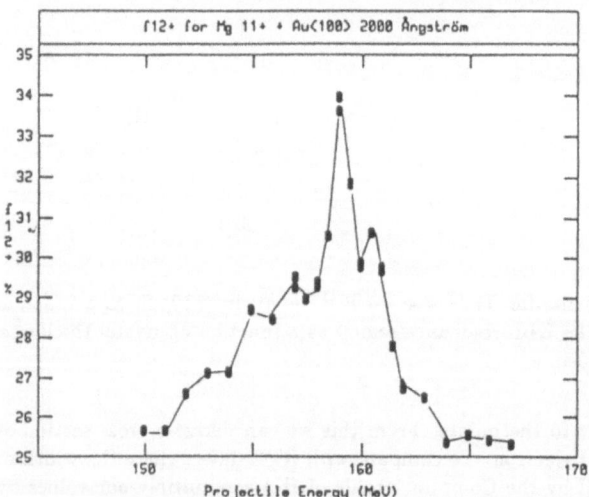

Figure 13

Yield of Mg^{11+} Ly α X-rays as from 1500 Å-thick Au $\langle 111 \rangle$ channeled Mg^{11+} ions as a function energy. The crosses and the rounded points represent two separate energy scans.

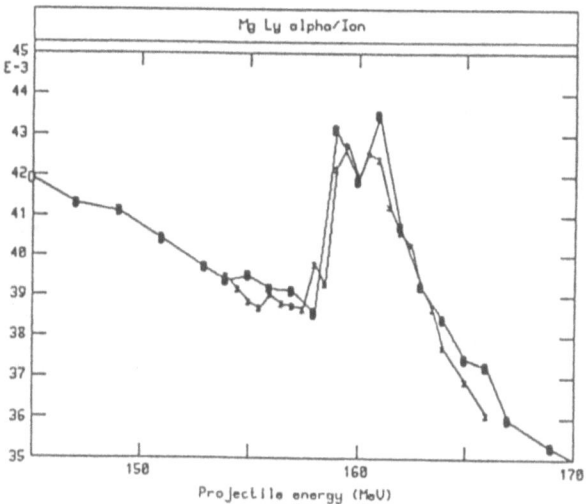

Figure 14

Charge fraction of Mg^{12+} emerging from an Au $\langle 111 \rangle$ channel 1500 Å-thick as a function of entering Mg^{11+} ion energy. Entrance slits to charge state analyser set to accept ±4.5 mr.

and the results for the charge state measurements are shown in Fig. 14. The seventh harmonic of the $n = 1$ to 2 resonance ($\Delta E_{ij} = 1468$ eV) should appear at 159.3 MeV if the transition were to occur in vacuum. The crystal field effects shift the $2p_{xy}$ peaks to lower energies and the $2p_z$ state to a higher energy. Thus, the upper of the two peaks corresponds to excitation into a $2p_z$ state and the lower to a $2p_x$ or $2p_y$ state. A comparison of the two curves shows that the probability of escape from the crystal without ionization (X-ray curve) is greater for ions in the $2p_z$ state than those in the $2p_{x,y}$ state.

Further experiments have shown the impact parameter dependence of both the degree of excitation and the relative probabilities of ionization.

Both the work on resonant coherent excitation and dielectronic excitation in channels show promise for enabling the study of electronic collisions of short lived excited states; processes which are of considerable importance in dense planmas such as those found in laser ablation and inertial fusion

This research was sponsored by the U.S. Department of Energy, Office of Basic Energy Sciences, Division of Chemical Sciences, under Contract No. DE-AC05-840R21400 with Martin Marietta Energy Systems, Inc.

References

1. S. Datz, C. Erginsoy, G. Leibfried and H.O. Lutz, Ann. Rev. Nucl. Sci. **17** 129 (1967).
2. D.S. Gemmel, Rev. Mod. Phys. **46** 129 (1974).
3. "Channeling", D.V. Morgan, ed., John Wiley & Son Ltd (1973).

4. S. Datz and C.D. Moak, "Heavy Ion Channeling" in "Treatise on Heavy Ion Science" Vol. 6, D.A. Bromley, ed., Plenum Press, 1984.

5. J.F. van der Veen, Surf. Sci. Rep. **5** 199 (1985).

6. "Surface Melting", Ph.D. thesis, A.W. Denier van der Gon, University of Leiden, 1990.

7. S. Datz, C.D. Moak, T.S. Noggle, B.R. Appleton and H.O. Lutz, Phys. Rev. **179** 315 (1969).

8. C.D. Moak, J. Gomez del Campo, J.A. Biggerstaff, S. Datz, P.F. Dittner, H.F. Krause, and P.D. Miller, Phys. Rev. **B25** 4406 (1982).

9. J.H. Barrett, B.R. Appleton, T.S. Noggle, C.D. Moak, S. Datz, J.A. Biggerstaff, and R. Behrisch, in "Atomic Collisions in Solids", S. Datz, B.R. Appleton and C.D. Moak, eds, Plenum Press (1975) pp. 645–688.

10. S. Datz, F.W. Martin, C.D. Moak, B.R. Appleton and L.B. Bridwell, Radiat. Effects. **12** 163 (1972).

11. S. Datz, C.R. Vane, P.F. Dittner, J.P. Giese, J. Gomez del Campo, N.L. Jones, H.F. Krause, P.D. Miller, M. Schulz, H. Schöne and T.M. Rosseel, Phys. Rev. Lett. **63** 742 (1989); see also Z. Phys. C. (in press).

12. B.R. Appleton, R.H. Ritchie, J.A. Biggerstaff, T.S. Noggle, S. Datz, C.D. Moak, H. Verbeek and V.N. Nelavathi, Phys. Rev. **B5** 2393 (1972).

13. F. Seely, Atomic Data and Nuclear Data Tables **26** 137 (1981).

14. W. Lotz, Z. Physik **206**, 206 (1968).

15. Y.-K. Kim and J.-P. Desclaux, Phys. Rev. **A38** 1805 (1988).

16. S. Datz, C.D. Moak, O.H. Crawford, H.F. Krause, P.F. Dittner, J. Gomez del Campo, J.A. Biggerstaff, P.D. Miller, P. Hvelplund and H. Knudsen, Phys. Rev. Lett **40** 943 (1978).

17. C.D. Moak, S. Datz, O.H. Crawford, H.F. Krause, P.F. Dittner, J. Gomez del Campo, J.A. Biggerstaff, P.D. Miller, P. Hvelplund and H. Knudsen, Phys. Rev. **A19** 977 (1979); Nucl. Inst. Meth. **194** 225 (1982).

18. P.D. Miller, H.F. Krause, J.A. Biggerstaff, O.H. Crawford, C.D. Moak, N. Nesković, P.L. Pepmiller and M.D. Brown, Nucl. Inst. Meth. **B13** 56 (1986).

19. S. Datz, P.F. Dittner, J. Gomez del Campo, K. Kimura, H.F. Krause, T.M. Rosseel, C.R. Vane, Y. Iwata, K. Komaki, Y. Yamazaki, F. Fujimoto and Y. Honda, Radiation Effects and Defects in Solids, 1991, in press.

The Frankfurt ECR-RFQ Ion-Beam Facility for Slow Highly Charged Ions

H. Schmidt-Böcking[1], *A. Schempp*[2], *and K.E. Stiebing*[1]

[1]Institut für Kernphysik, Johann Wolfgang-Goethe Universität,
  W-6000 Frankfurt am Main, Fed. Rep. of Germany
[2]Institut für angewandte Physik, Johann Wolfgang-Goethe Universität,
  W-6000 Frankfurt am Main, Fed. Rep. of Germany

1   Introduction

Very recent developments in the technology of ion sources for highly charged ions[1-23], the acceleration[24-27] and storage[29,30] of such ions has initiated substantial world-wide physics activities with slow highly charged ions and in particular studies of the interactions of these ions with atomic and solid state targets. Due to their high charge states, they carry a strong ionizing power into collisions with atoms even in cases where their kinetic energy is too small to induce electronic ionization and excitation through collision dynamics. The highly charged ion itself bears an amount of electronic potential energy sufficient to induce numerous characteristic reactions in a collision with a single atom or with the atoms of the surface of a solid. These characteristic reactions can be used for the modification or the analytical characterisation of the target system in general and of a solid surface in particular.

In the last few years substantial effort has been made to better understand the atomic physics of the interaction of a slow highly charged ion with different target systems and to use such beams for materials research. In many cases, due to technical reasons, research centers have been established at laboratories where ECR-ion sources existed as part of the beam injection system for high-energy nuclear-physics research accelerators. However, the beam energies available for highly charged ions from these sources are either very low ($E_{ion} \leq 1.\text{keV/u}$) or rather high ($E_{ion} \geq 1.\text{MeV/u}$). Only very few facilities [12,16] could deliver beams with variable energies (up to 20keV/u for uranium beams). Most of these facilities produce high energy beams for use in nuclear physics experiments. Therefore only to a small percentage of their time is available for materials research and materials modification. For this reason there is a strong demand for additional facilities delivering intense beams of highly charged heavy ions for atomic physics and materials research. Those facilities should cover an ion-beam velocity range

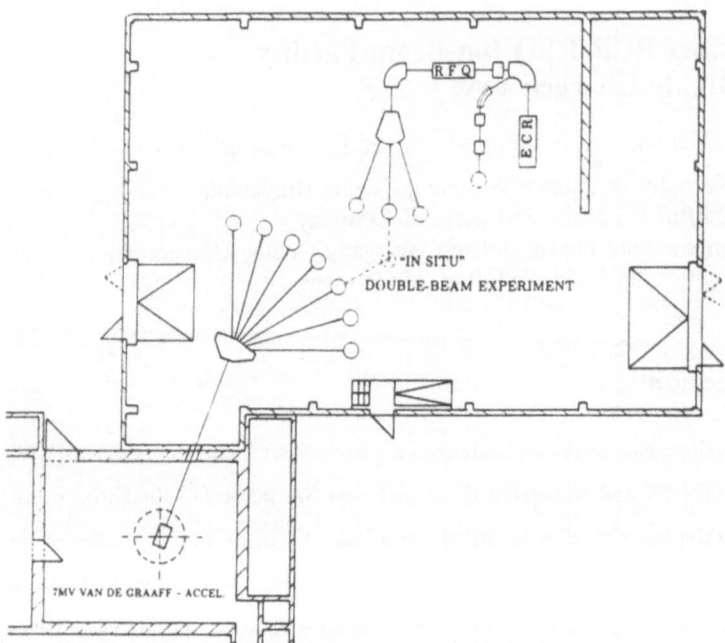

Fig.1: The experimental area at the Institut für Kernphysik, University of Frankfurt, FRG with the ECR-RFQ-facility and beam lines from the 7MV van de Graaff accelerator.

where the variable ion beam velocity can be tuned to be much smaller and much larger than the relevant electron velocities in the outer atomic states. Thus a study of quasistatically and dynamically induced electronic transitions is possible. The ion-velocities available should therefore range from $v_{ion} \approx 0.1$a.u. up to $v_{ion} \approx 3$.a.u..

The new ECR-RFQ facility presently under construction at the University of Frankfurt, Germany will provide such beams. Its basic components are:

- a 14.6 GHz ECR-ion source based on the design of the Berkeley AECR [7] and

- a compact RFQ-accelerator with variable final energy, which has been designed and is presently built at the Institut für Angewandte Physik at the University of Frankfurt [25-27].

Furthermore, the system will include devices to generate pulsed beams that can be adjusted over a wide range of time structures depending on the demands of the ex-

72

periments to, e.g., allow the employment of time of flight ion detection techniques. In Fig.1 a scheme of the whole facility is shown. The system will be mounted in the new experimental area of the 7 MV van de Graaff accelerator. The compact facility will be connected to the van de Graaff beam lines allowing in-situ double-beam experiments.

In section 2 the design of the ECR-ion source is discussed. In section 3 the RFQ-accelerator and the beam pulsing system is described. In section 4 a survey of some of the atomic-physics and materials-research experiments is given.

2   The ECR– Ion Source

The 14.6 GHz ECR source is designed and constructed following the concept of the LBL-AECR ion source [6]. The frequency of 14.6 GHz has been chosen to obtain reasonable intensities of the highest charge states without the necessity to install a superconducting magnet device. In Fig. 2 the scheme of the designed ECR source is shown. Since a

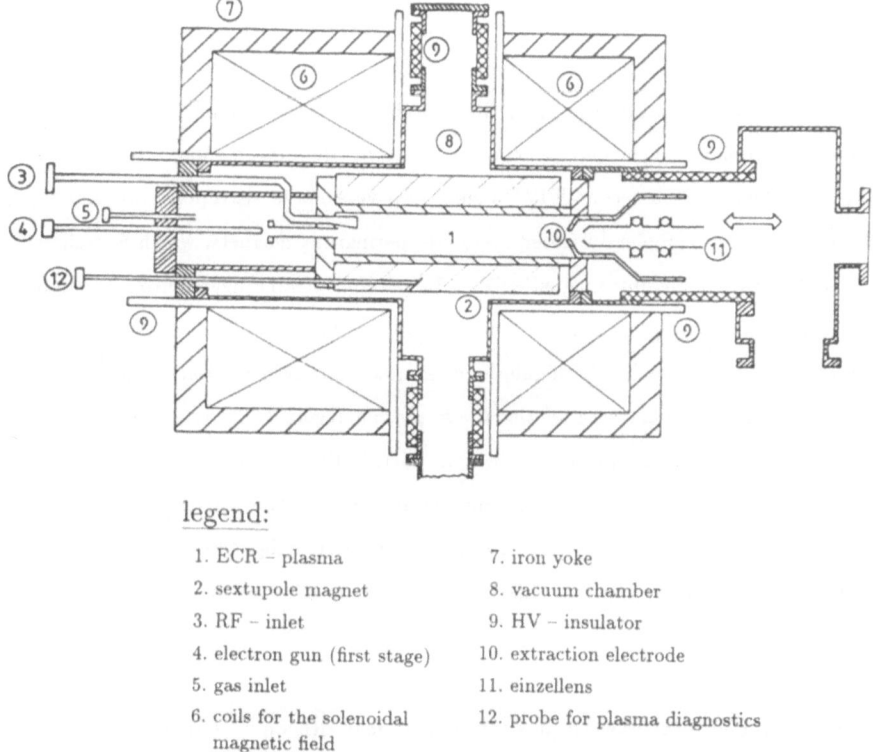

legend:

1. ECR – plasma   7. iron yoke
2. sextupole magnet   8. vacuum chamber
3. RF – inlet   9. HV – insulator
4. electron gun (first stage)   10. extraction electrode
5. gas inlet   11. einzellens
6. coils for the solenoidal   12. probe for plasma diagnostics
   magnetic field

Fig.2: Delineation of the ECR-ion source for the Frankfurt ECR-RFQ facility .

wide variation of the beam velocity is important for the desired research projects, the source is designed to yield highest terminal voltages without mounting the source on a separate high voltage platform. The design is made in such a way that the high voltage and ground potential parts are separated by a very simply structured insulation. This insulation part consists of two circular solid pieces, which provide up to 100 kV insulation between terminal and ground potential.

The source vacuum of about 10^{-6}mbar in the main plasma region is obtained by a 1000 l/sec diffusion pump connected to the terminal part via a 15cm long ceramic insulator.

The magnetic trap is formed in the usual way by the superposition of a solenoidal magnetic field and the field of a magnetic sextupole. The solenoidal field is generated by three coil packages consisting of a total of 22 "pancake"–coils. These three packages have separate power supplies. The whole magnetic system is surrounded by an iron yoke to minimize stray fields. The total iron yoke system is constructed to allow for modification of the actual magnetic field (movable iron rings in the vacuum on both sides of the plasma) thus enabling necessary correction to the solenoidal fields in the plasma region of the first ECR–stage and in the extraction area of the source. The sextupole is situated inside the main vacuum chamber. It is completely sealed by a thin copper tubing, which serves as cooling device for the magnets and also forms the plasma chamber of the second ERC-stage. Two versions of sextupoles are planned. One is a compact, completely closed system of permanent magnets, which is designed to yield the highest possible magnetic field . Its design is nearly identical to a sextupole installed in the Groningen ECRIS source [13]. A second sextupole consists of separated bars of permanent magnets. It therefore has a somewhat lower magnetic field, but will allow spectroscopic diagnostics of the whole plasma region to be carried out. With this sextupole the plasma properties (density distributions, plasma cooling, afterglow effects etc) will be studied. The sextupole is mounted on the entrance flange of the source. All necessary feed throughs for cooling, the spectroscopic probes, gas inlet and RF feed-through are arranged on this flange in order to allow a fast and easy opening of the source system.

The construction of the source allows the mounting of either a 'traditional' first stage (separately RF powered first plasma stage) or an electron gun as the injection system

[7]. For the beam extraction an einzellens system is presently designed. For the beam transport focussing electrostatic quadrupole lenses (doublets and triplets) will be used.

3 The RFQ–accelerator

Radio Frequency Quadrupole (RFQ) structures are very attractive for low energy ion accelerators. They cannot replace static injectors and Van de Graaff generators in terms of energy resolution and beam quality but are favourable for applications with high current beams or in combination with sources like an ECR because the source can be close to ground potential and is easy to operate and service.

The RFQ concept of spatially homogeneous strong focusing proposed by Kapchinskij and Teplyakov[24] employs strong focusing with rf-electrical focusing which is independent on velocity. Therefore the accelerator can start at low energy with short cells. This allows an adiabatic capture of the dc beam from the ion source to a bunched beam with a high transmission of up to 100% with a small emittance growth. The RFQ basically is a homogeneous transport channel with additional acceleration. The mechanical modulation of the electrodes as indicated in Fig.3 adds an accelerating axial field component, resulting in a linac structure which accelerates and focuses with the same rf fields.

The focusing gradients $G \sim X \times U_Q/a^2$ ($X < 1$ for modulated electrodes) determine the acceptance for a given injection energy and frequency in a low current application. A maximum voltage U_Q has to be applied at a minimum beam aperture a, if the radial focusing strength is the limiting factor. The highest possible operation frequency should be chosen to keep the structure short and compact. The choice of U_Q, the operating frequency, the values of aperture a, modulation m and the lengths L_c along the RFQ, determine the electrode shape (pole tips) and the beam properties. The design of the RFQ has to be made for the heaviest particle to be accelerated and for fixed initial and final ion velocities v_p (or specific energy T/u) because of the velocity profile given by the geometry of the electrodes and the Wideröe resonance condition: $L_i = \beta_p \lambda_o/2 = v_p/2f$. The operating frequency of the four-rod RFQ is changed to vary the ion velocity v_p or the energy T: $v_p \sim f, T \sim A/q \times f^2$. Acceleration of ions with mass A and charge state q to lower energies requires a lower electrode voltage $U_Q \sim A/q \times f^2$.

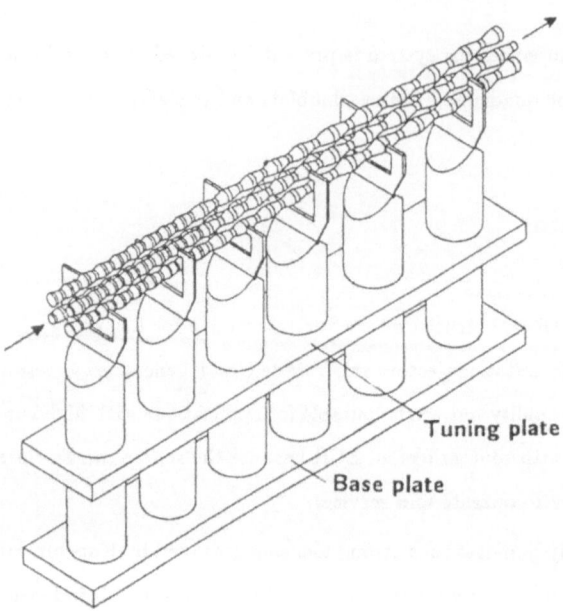

Fig.3: The principle of the variable-energy (VE) RFQ

To change the frequency of the 4-Rod-RFQ the resonator is tuned inductively by changing the effective length of the driving conductor with movable shorts as indicated in Fig.3. The VE-RFQ, built for the ECR-RFQ combination at the IKF at the University of Frankfurt [25,26] has a frequency variation of f= 80-110 MHz for which a FM-transmitter has been rebuilt. It is designed for a minimum specific charge of q/A=0.15, an output ion energy of T_f=100–200 keV/u, an electrode voltage of U_Q= 70 kV and has a structure length of L_s=1.5m. The acceptance of the RFQ is calculated to $\alpha_N = 0.5\pi \cdot mm \cdot mrad$ and the emittance growth is 50%. Fig.4 shows the electrode parameters along the RFQ.

The RFQ electrodes have to be periodically charged by a resonator as rf current source. The 4-Rod resonator consists of coupled $\lambda/2$ oscillators in a linear arrangement and, although the current densities at the electrode supports are relativly high, the efficiency is very good. An rf-power of 50 kW (25% duty cycle) is needed for the highest energy and frequency.

A special feature of this ECR-RFQ combination is the chopper system, which is planned to deliver single ion bunches to the experiment. A chopper in front of the RFQ delivers pulses approximately 1μs long to the RFQ. From this pulse train an anharmonic

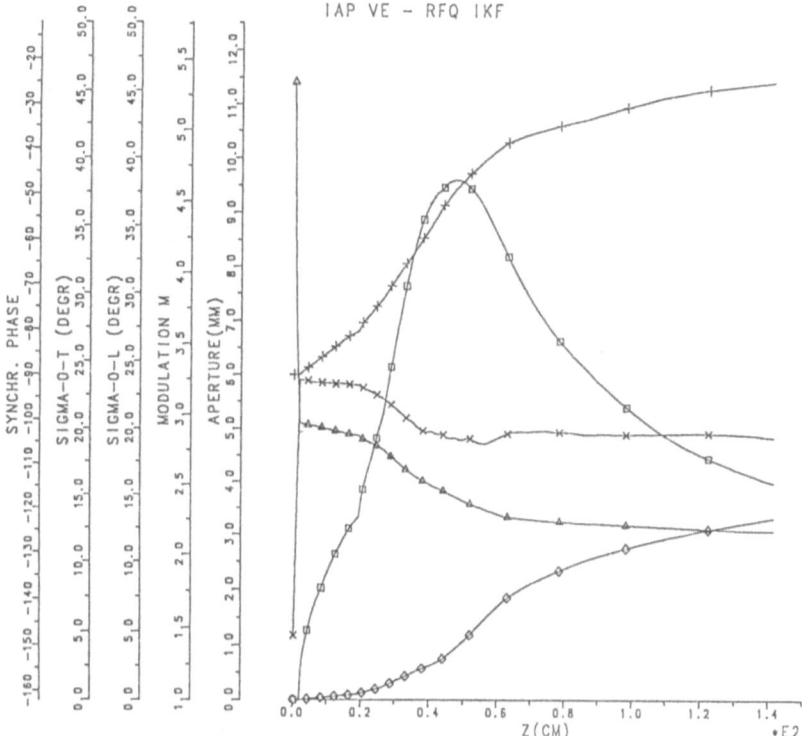

IAP VE – RFQ IKF

Fig.4: The electrode parameters along the RFQ

chopper following the the RFQ can then select a single bunch to be transported to the experiment.[28]

4 Research program

4.1 Atomic physics

The electron transfer between highly charged ions and rare gases, molecules and solid targets will be systematically investigated. The understanding of this charge transfer is of fundamental interest for the improvement of many particle charge transfer theory, but also for the application of such ions in materials research[31-41]. Particularly the inner shell charge transfer as function of the impact parameter b will be studied as function of the initial and final charge states. This inner shell transfer obviously plays a very important role, when swift highly charged ions interact with solid state targets. The analysis of Auger-electron emission indicates that those processes contribute substan-

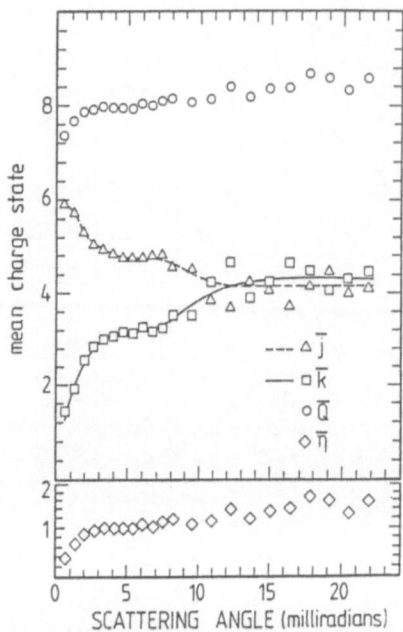

Fig.5: Mean recoil charge states $\bar{k}(\Theta)$, mean projectile charge state $\bar{j}(\Theta)$, and mean total charge $\bar{Q}(\Theta)$. $\bar{\eta}$ is the mean number of electrons emitted per collision by autoionization.

tially to the so called "side feeding" channel [31]. In a first systematic study of charge transfer between 90 keV Ne^{7+} and Ne^{9+} ions with the Ne [41] we have measured, by recoil-projectile coincidence techniques, the angular dependence of the many electron transfer. The detailed analysis of all data measured shows that dynamically induced target or projectile ionization in this low velocity regime is extremely unlikely and is not detectable in the data. Autoionization of doubly/multiply excited ions, however, is a very likely event as soon as two electrons are shared between both partners. The probability of sharing two or more electrons is close to 100% for impact parameters b smaller than about 2 atomic units. In Fig.5 the mean charge states $(\hat{j}, \hat{k})$ for target and projectile ion respectively as functions of the projectile scattering angle Θ are shown for Ne^{7+} on Ne. In the lower part of the figure the contribution of autoionization is plotted. For larger angles the number of autoionization electrons per collision is larger than one. For $\Theta < 10\,\text{mrad}$ only the projectile autoionizes, because it captures the electrons into high lying levels whereas the target remains in the ground state. For $\Theta > 10\,\text{mrad}$ also the target autoionizes, i.e. it has vacancies in its L-shell. This sudden opening of the target autoionization channel at $\Theta \approx 10$mrad is due to the transfer of target L-electrons into the projectile L-shell and allows the determination of the cross section of this process. We obtain cross sections of more than 1 atomic unit ($\cong 10^{-16}$cm^2), indicating the importance of the inner shell charge transfer. One should notice, however, that in a more asymmetric collision this transfer cross section may be reduced since the charge transfer probability is decreased because of increased level splitting. From the 90keV Ne^{9+}-Ne collision experiment we obtain the cross section for the direct K-K-transfer. This cross section is also large and nearly equal to the geometrical cross section of the quasimolecular 2pσ–1sσ radial coupling region ($\Delta\sigma_{K-K} \approx 0.1$a.u.). It is of interest that the measured Ne^{9+}-Ne many-electron charge transfer probabilities can well be described by two independent processes (i) the resonant K-K-transfer process and (ii) the outer shell transfer process as measured for Ne^{7+} on Ne.

The measurement of coincidences between charge state resolved projectiles and recoil ions is not sufficient to resolve the n,l-value of the captured state. Therefore electron-recoil/projectile coincidence experiments are needed to get more detailed information on the n,l- population. Such electron/x-ray-particle coincidences have been performed to study the quasimolecular radiation emission. Such measurements provide information of the inner shell population during the collision. From measurements of the 1sσ-x-ray emission at small impact parameters in Ne^{9+}-Ne-collisions we deduced that the quasimolecular inner shell states are nearly empty during the collision ("hollow" quasi-molecule) and are filled by quasimolecular promotion on the outgoing part of the collisions. The investigation of the quasimolecular x-ray and electron emission as functions of polar and azimuthal scattering angle will be a central goal of our research program. The detection devices and the multiparameter data acquisition systems have already been developed.

4.2   Materials research

The interaction of highly charged slow and swift ions with solid targets and surfaces will be another major research project at the new facility. At several other highly charged ion source facilities, charge exchange at the surface has been studied by electron and x-ray spectroscopy and detailed information on the neutralization process including cascading decay processes in the highly charged ion has been obtained [42-46]. We are planning to extend such measurement to higher velocities, where full neutralization of the ion in the surface region cannot be reached. Ions backscattered from the surface may still remain in a high charge state. Charge exchange along well defined trajectories will be studied, using oriented crystals as targets, partially in coincidence with the emitted electrons. An apparatus is under construction with which charge exchange under surface channelling conditions can be investigated. In addition, the charge exchange in the solid near the surface will be studied. To get first qualitative information on the filling of the empty projectile shell when a slow highly charged ion penetrates a surface and the solid, the stopping power of highly charged 150 keV Ar ions as function of the incoming charge state has been measured in 120 Å thick C foils[47]. It could be shown that the highly charged Ar ions must capture several electrons into its L-shell in the first layer. Thus the ion nearly immediately reaches the mean charge state corresponding to this ion velocity. The stopping power measured appeared independent of the initial charge state. Only the rate of electron and photon emission on the entrance side of the foil was strongly charge state dependent. The rates on the exit side showed no incoming charge state dependence. Also range profiles for 150 keV Ar^{9+} in Si have been measured and no dependence on the initial charge state has been found.

Since the facility allows a wide variation of the beam pulsing structure material modification can be studied for any time structure ($> 10^{-8}$sec) of the implantation rate. The variable beam pulsing structure enables time-of-flight measurement of surface sputtered ions and of neutral reaction products. One beam line after the RFQ will be prepared in such a way that the highly charge ion bunches can be focussed and collimated to form a μ-beam allowing a highly position resolved surface analysis or modification. Dependent on its charge state the highly charged ion carries a huge amount of potential energy in its empty shell which will be released immediatly when the ion touches the surface, inducing ionization and exitation of the solid material. All target atoms near the impact region ($\geq$10Å) should be ionized. Thus we may obtain a very controlled ionization of target material in the sub μm regime. The beam line for the μ-beam will be designed and mounted when in 1993 the whole facility is successfully in operation.

Acknowledgements

We are indebted to Prof. K. Ring, Dr. Biermann, Dr. R. Maas, R. Dueker, Dr.F. Dietz, Prof. Dr. J.Bereiter-Hahn, Dr. Begemann and many other in the University administration, the Ministerium für Wissenschaft und Kunst des Landes Hessen, and the DFG for their continuing support to get this project funded. In particular we are thankful for numerous very stimulating and helpful discussions with our colleques and friends

K. Lyneis, M. Prior, C.L. Cocke, A. Drentje, R.Morgenstern, J. Andrae, F. Meyer,

T.A. Antaya, R. Geller, H. Beuscher, P. Sortais, E. Salzborn, M. Liehr, H. Büttig,

G. Zschornack, H. Streitz, K. Bethge, H. Klein, N. Angert, P. Mokler, J. Ullrich,

H. Thyrroff, J. Beyreuther, R. Stock, H. Deitinghoff, R. Becker, J. Friedrich,

J. Klabunde, D. Hoffmann, R. Herrmann, K. Wisemann and B. Huber.

References

[1] R. Geller, Nucl. Sci. NS23 (2), 904 (1976)

[2] R. Geller, Nucl. Sci. NS26 (2), 2120 (1979)

[3] R. Geller, Proceed. of the 5th Int. Conf. on the Physics of Highly Charged Ions, Gießen, Germany, ed. by E. Salzborn, P.H. Mokler, A. Müller, Z. Phys. D21, 117 (1991)

[4] R. Geller, B. Jaquot and P. Sortais, Nucl. Instr. and Meth. A243, 244 (1986)

[5] G. Melin, F. Bourg, P. Briand, J. Debernardi, M. Delaunay, R. Geller, B. Jaquot, P. Ludwig, T.K. N'Guyen, M. Pontonnier and P . Sortais, Journal de Physique Coll.C1, suppl. au no 1, C1-6723 (1989)

[6] C. M. Lyneis, in ref. [5], p. C1-698

[7] C. M. Lyneis, Zuqi Xie, D.J. Clark, R.S. Lam, S.A. Lundgren, Proceed. of the 10th Int. Workshop on ECR Ion Sources, Knoxville, Tenn., ed. by F.W. Meyer and M.I. Kirkpatrick, 47 (1991)

[8] T.A. Antaya, in ref [5], p. C1-707

[9] P. Sortais, P. Attal, L. Bex, M. Bisch, M.P. Bourgeral, Y. Bourgoin, P. Lehrissier, J.Y. Pacquet, in ref.[5], p. C1-855

[10] H. Beuscher, in ref.[5], p. C1-883

[11] M. Arnould, F. Baeten, D. Darquennes, Th. Delbar, C. Dom, M. Hyse, Y. Jongen, M. Lacroix, P. Leleux, P. Lipnik, M. Leuiselet, G. Reusen, G. Ryckewart, Sindano Wa Kitwanga, P. van Duppen, J. Vanhorenbeeck, J. Vervier, S. Zaremba, in ref. [5], p. C1-813

[12] R.C. Pardo, P.J. Billquist and J.E. Dey, in ref.[5], p. C1 695

[13] A.G. Drentje and J. Sijbring,in ref.[7], p. 17

[14] M. Liehr, M. Schlapp and E. Salzborn, in ref.[7], p. 363

[15] F.W. Meyer, R.A. Phaneuf, D.C. Gregory, C.C. Havener, J.W. Hale, P.A. Zeijlmans van Emmichoven and J.S. Thompson, in ref.[7], p. 367

[16] M.P. Stöckli, in ref.[3], p. 111

[17] E.D. Donets Inventor's Certificate No. 248860, March 16,1967 Byull, Oipotz No. 23, 65 (1969); E.D. Donets, G.D. Shirkov, Avtorskoe Svidetelsvo USSR, N1225420 (1984), Bul. OI,N44, p.69 1989;

E.D. Donets, "The Physics and Technology of Ion Sources",
ed. by I.G. Brown (John Wiley, New York), 245 1989;
V.P. Ovsyannikov, 5th Int. Conf. on the Physics of Highly Charged Ions,
Gießen, Germany, Book of Abstracts, 5.13 1990;
E.D. Donets, Int. Conf. on Ion Sources, Berkeley, CA (1989)

[18] M.A. Levine, R.E. Marrs, J.R. Henderson, D.A. Knapp and M.B. Schneider,
Proceed. of Workshop and Symposium on the Physics of Low-Energy Stored and
Trapped Particles, Stockholm, Sweden (1987), Physica Scripta T22, 157 (1988)

[19] R.E. Marrs, C. Bennett, M.H. Chen, T. Cowan, D. Dietrich, J.R. Henderson,
D.A. Knapp, M.A. Levine, K.J. Reed, M.B. Schneider and J.H. Scofield,
in ref.[5], p. C1-445

[20] J. Arianer, A. Caberspine, C. Goldstein, Nucl. Instr. and Meth. 193 (1982), 401

[21] S. Ohtani, Physica Scripta T3, 110 (1983)

[22] M. Kleinod, R. Becker, H. Klein, Int. Symp. on Electron Beam Ion Sources
and the Applications, Upton, NY (1988) AIP Conf. Brok. NO 188
(American Institute of Physics, New York, 1989)

[23] M.A. Levine, R.E. Marrs, C.L. Bennett, J.R. Henderson, D.A. Knapp
and M.B. Schneider, in ref.[22], p. 82

[24] I.M. Kapchinskij, V.A. Tepliakov, Prib. Tekh. Eksp. 4, 17 , 19 (1970)

[25] A. Schempp, Linac88, Cebaf80-001, 460 (1989)

[26] A. Schempp, Nucl. Instr. and Meth. B40/41, 937 (1989)

[27] A. Schempp, Nucl. Instr. and Meth. B50, 460 (1990)

[28] A. Schempp, SLAC-Report 303, 257 (1986)

[29] P. Hvelplund, in ref.[5], p. C1-459

[30] B. Franzke, in. ref.[18], p. 41 and other papers therein

[31] R. Morgenstern, this book,

[32] A. Müller, in ref.[3], p. 39

[33] S. Datz, P.F. Dittner, J. Gomez del Campo, H.F. Krause, T.M. Rosseel and
C.R. Vane, in ref.[3], p. 45

[34] L.H. Andersen, in ref.[3], p. 29

[35] A. Barany and H. Danared, Proceed. of the Conf. on the Physics of Multiple
Charged Ions, Groningen, Netherlands (1986),
Nucl. Instr. and Meth. B23,1 (1987)

[36] R.K. Yanev, in ref.[5], p. C1-421

[37] R.K. Yanev and H. Winter, Phys. Rep. 1170, 267 (1983)

[38] B.A. Huber, in ref.[35], p. 34

[39] C.L. Cocke, in ref.[5], p. C1-19

[40] C.C. Havener, M.S. Huq, F.W. Meyer and R.A. Paneuf, in ref.[5], p. C1-7

[41] H. Schmidt-Böcking, M.H. Prior, R. Dörner, H. Berg, J.O.K. Pedersen, C.L. Cocke,
M. Stockli and A.S. Schlachter, Phys Rev. A37, 4640 (1988)

[42] J.P. Briand, L. de Billy, P. Charles, J.P. Desclaux, P. Briand, R. Geller, S. Bliman
and C. Ristori, in ref.[3], p. 123

[43] E.S. Parilis, in ref.[3], p. 127

[44] H. Winter, in ref.[3], p. 129

[45] H.J. Andrä, A. Simijonovici, T. Lamy, A. Brenac, G. Lamboley, J.J. Bonnet,
A. Fleury, M. Bonnefoy, M. Chassevent, S. Andriamonje and A. Pesnelle,
in ref.[3], p. 135

[46] F.W. Meyer, C.C. Havener, S.H. Overbury, K.J. Snowdon, D.M. Zehner, W. Heiland and H. Hemme, in ref.[35], p. 234

[47] R. Herrmann, C. Neelmeijer, R. Grötschel, H. Sharabati, H. Beuscher, H. Schmidt-Böcking, proceedings of the 2nd European Conf. on Accelerators in Applied Research and Technology (ECAART 1991), to be published in Nucl Instr. and Methods $\underline{B}$

Role of Ion Beams in Superconductor Research

O. Meyer

Kernforschungszentrum Karlsruhe,
Institut für Nukleare Festkörperphysik, P.O. Box 3640,
W-7500 Karlsruhe, Fed. Rep. of Germany

Ion Beams have contributed to high-T_c superconducting research mainly in four ways.
(i) Rutherford backscattering and channeling is widely used for the composition and growth analysis of epitaxial thin film synthesis. (ii) Ion channeling has further been used to investigate the structure and phonon properties of HTSC-single crystals. (iii) Ion beams are used to perform fundamental defect and irradiation effect studies and (iiii) to modify superconducting properties in a controlled fashion for application. The results discussed in this review indicate a substantial contribution of ion beams to the rapid progress in high-T_c research.

1. Introduction

Ion backscattering and channeling spectrometry is a well established technique for the analysis of thin films and material surfaces [1, 2]. From the beginning of HTSC thin film deposition RBS analysis proved to be a very useful method to optimize the deposition parameters with respect to film composition and lateral homogeneity [3]. When epitaxial film growth was realized, channeling became a powerful tool to analyse the structure and disorder at surfaces, interfaces and in the bulk. Energy dependent channeling measurements reveal information about defect structures present in films and single crystals. The channeling method combined with nuclear reactions or resonant scattering can be used to analyse the oxygen sublattice. This is very important because the oxygen concentration and lattice site occupation is known to play a dominant role for the properties of HTSC. Above that channeling is applied to optimize the parameters for substrate surface and bulk quality improvement. A perfect substrate surface finish is an essential prerequisite for best film growth. In this way deposition parameters could be optimized to achieve high quality epitaxial thin film growth [3]

Ion channeling in high-quality high T_c-single crystals has revealed the existence of anomalous phonon behavior at T_c, providing information on the electron-pairing mechanism [4]. The channeling studies allowed to measure changes of the atomic displacement amplitude smaller than 1pm. The presence of twins as well as the existence of small, periodic lattice displacements and incommenserate phase modulation was also studied by ion channeling [5, 6].

The superconducting properties of materials are strongly affected by chemical and structural disorder. Ion beams are used to modify the materials and study such effects in a controllable fashion either by producing radiation induced disorder (in this case the ions penetrate the material to be studied) or by doping (the ions are stopped and the composition of the material is varied). The influence of disorder range up to levels where phase transformations e.g. amorphization will occur. The irradiation with fast ions in the energy range of a few keV to MeV is a well established technique to introduce defects in materials by atomic collisions in a quantitative and reproducible manner. In high-T_c superconductors irradiation experiments at different temperatures mostly performed on thin films deliver interesting results with respect to their practical applications and in basic science. For a review covering these results in more detail and or further references see [7].

HTSC are rather sensitive to irradiation induced disorder. The T_c depression rate, however, is strongly depending on the initial growth quality of the material characterized, e.g., by its residual resistivity, ρ, residual resistivity ratio, and grain boundary density [8]. Better growth results in less sensitivity to irradiation and good material reveals a sensitivity situated between that of the Chevrel and the A15 phases. This result may encourage the application of HTSC in radiation environment, especially because beneficial effects have been found concerning the critical current density J_c in irradiated materials at high magnetic fields [9, 10]. The control of J_c was also used to optimize the sensitivity of superconducting devices [11].

The structure of defects is of large interest in HTSC. Significant recovery effects with respect to ρ and T_c after low temperature irradiation [12] and an increasing c-axis lattice parameter with irradiation fluence [13] indicate that displaced oxygen atoms play an important role for the observed property changes. This underlines the importance of proper oxygen incorporation in HTSC-material preparation and especially thin film growth.

The resistivity of irradiated HTSC films increases superlinearly with fluence [14] finally leading to a metal to insulator transition which structurally is accompanied by total amorphization of a sample [15]. While the superlinear behavior and the amorphization mechanisms are of basic interest, the result of insulation is practically exploited for fully planar patterning of thin films. Thus, narrow bridges for critical current measurements and the first SQUID have been fabricated using ion beams [16].

In the following we present a review mainly of our own results on the application of ion beams in the growth analysis of thin epitaxial films and on property modifications of thin films by ion irradiation.

2. Epitaxial growth of thin high-T_c films

Thin films of the high-T_c superconductors have been the subject of many studies as they are of great interest for basic research as well as for technology. Thin films are used for critical current measurements, for tunneling probe and optical investigations, and for studies concerning the effect of ion induced modifications of the physical properties. Applications would include Josephson junctions for computer circuits, SQUIDs, high-frequency transmission lines, interconnects and other passive and active electronic devices. In science and technology epitaxial films on various substrates are wanted in order to fulfill all the requirements imposed by special needs.

Especially for high T_c material it became soon apparent that high J_c values in the range of 10^6 A/cm^2 are closely related to the growth quality of the films [17]. In general it turned out that polycrystalline films in contrast to different structure classes of the "classical" superconductors revealed degraded normal-conducting and superconducting properties. This probably is due to a granular nature of such films where high-T_c grains are surrounded by deteriorated material which may be oxygen deficient. Films having such an inhomogeneous structure are also less suited for basic research experiments, like e.g. for irradiation studies where the investigation of defect structures is of interest, or for application in electronic device technology where high potentialities are expected for the use of the new oxide superconductors. Therefore all attempts in thin film preparation concentrated on the deposition of highly oriented, textured films with the aim of achieving single-crystalline growth in order to fulfill all the requirements imposed by special needs.

Quite a number of different deposition techniques have been applied for the preparation of thin films of the superconducting oxides. For basic studies of the epitaxial growth a reliable and reproducible technique is required such that the structure and the properties of the films can be studied as a function of the deposition parameters which are essential for epitaxial growth. Besides the deposition parameters the choice of the substrate material is important for growth studies for two reasons: first, at the substrate temperatures applied during deposition, substrate-film reactions should be avoided, and second, a good lattice matching between the growing film and the substrate surface which is beneficial for epitaxial growth should be achieved. A high quality substrate surface finish is another prerequisite for epitaxial growth.

Different methods may be applied for the structural characterization of the growth quality of epitaxial films. With respect to diffraction methods X-ray diffraction is most commonly used as a fast and simple technique because the use of reflection geometries does not require special sample preparation. The growth direction of films, the mosaic spread or admixtures of grains with different crystallographic orientations are easily determined from diffraction experiments. The drawback of X-ray measurements is their relative insensitivity to intrinsic defects and the lack of depth resolution. These drawbacks are compensated by ion channeling combined with Rutherford backscattering or nuclear reactions. This techniques is not only very sensitive to intrinsic deviations from perfect growth but also provides depth resolution allowing special studies of interface or surface effects. A very powerful technique for the growth characterization also is high-resolution electron microscopy which, however, needs special sample preparation and therefore cannot be applied non-destructively.

2.1 Substrate surface analysis

The amount of polycrystalline material in highly textured films depends on the substrate surface quality and on the substrate temperature and orientation. The influence of the substrate surface quality on the film growth performance is demonstrated in fig. 1 [17]. In fig. 1a random and [100]-aligned spectra of (100) $SrTiO_3$ substrates are shown. The sample providing the aligned spectrum (a) with $\chi_{min} = 22$ % was cut into three pieces which were annealed at 950^0C in 1 atm O_2 for 1h, leading to an improvement of the crystalline quality shown as spectrum (b) with $\chi_{min} = 9$ %, and at 1300^0C shown as spectrum (c) with $\chi_{min} = 2$ %. The Sr surface peak areas decreased correspondingly. YBaCuO films were then deposited under optimized conditions resulting in c-axis oriented growth. In fig. 1b the backscattering spectra of these films which are labelled according to their corresponding substrates are displayed. The random spectrum is taken from film (c); it is similar for all the films except of small differences in the film thickness. The χ_{min}values vary from 43 to 33 % and further to 15 % for the films on the substrates of increasing quality. Since similar χ_{min}values were reproducible obtained for the films on the substrates with similar quality, these results clearly demonstrate a strong dependence of the film growth on the substrate quality. X-ray diffraction measurements in the Seemann-Bohlin geometry reveal polycrystalline phase contributions for the films (a) and (b) while for the film (c) no lines could be detected indicating epitaxial growth.

Besides $SrTiO_3$, a number of substrates like Al_2O_3, MgO, ZrO_2, $LaAlO_3$ and $LaGaO_3$ have been analysed (3, 18, 19). In general it was observed that disordered substrate surfaces could be strongly improved by annealing in 1 atm O_2 at high temperature.

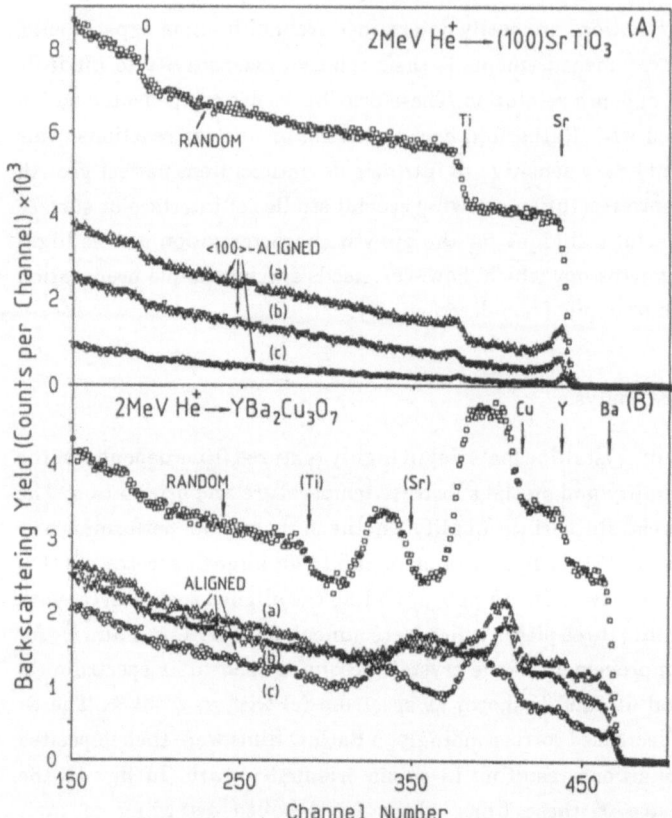

Fig. 1   Random and [100]-aligned backscattering spectra from (A) (100) SrTiO$_3$
          substrates of various qualities (χ_{min} is (a) 22 %, (b) 9 %, and (c) 2 %) and
          (B) c-axis oriented YBaCuO films deposited on them (χ_{min} for Ba is (a)
          43 %, (b) 33 %, and (c) 15 %).

2.2 Films

The crystalline quality of thin YBaCuO films should be compared to that of
bulk single crystals. Minimum yield values of 3.5 % (20) and 2 to 3 % (5) have been
measured by MeV He ion channeling for the Ba-sublattice of YBa$_2$Cu$_3$O$_{7-\delta}$ single
crystals. The values are still larger than a value of 1.5 % which is expected for a
perfect YBa$_2$Cu$_3$O$_7$ single crystal from Monte Carlo simulation (3). The best
minimum yield values observed up to now for YBaCuO thin films on different
substrates are 3 %, 3 %, 6 %, and 5% for films on (100) LaAlO$_3$, (100) SrTiO$_3$, (100)
MgO, and 100 Zr(Y)O$_2$, respectively. As an example the random and aligned
spectrum for a 3000 Å thick YBaCuO thin film deposited by magnetron sputtering
on (100) SrTiO$_3$ are shown in fig. 2. A value of 3 % has been obtained indicating a
crystalline quality similar to that of bulk single crystals. These films had a zero

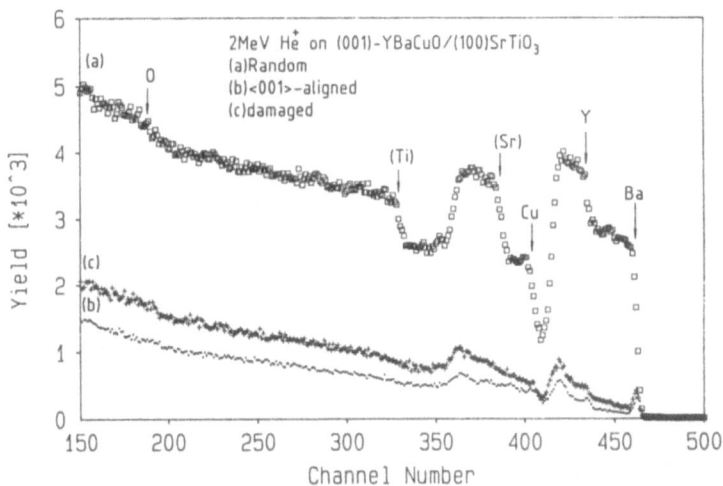

Fig. 2   Random and [100]-aligned backscattering spectra from a c-axis oriented
         YBaCuO thin film deposited on (100) SrTiO$_3$ substrates.

resistance temperature of 91 K, which J_c (77 K) $\geq 5 \times 10^6$ A/cm^2. The microwave
properties of these films consisted of surface resistance $R_s < 0.5$ m Ω at 6 GHz and ~
1 m Ω at 10 GHz, results which are one order of magnitude better than that of
copper at 77 K. From SEM, HRTEM and inclined channeling measurements it is
known that the films are highly twinned and contain a large number of planar
defects. From the Ba surface peak as seen in the aligned spectrum of fig. 2 the
thickness of the disordered surface layer can be estimated. For this analysis the
contribution of the first Ba-layer and the thermal contribution from successive Ba
atoms in the Ba$_2$Y row have to be subtracted from the measured peak area. For the
best surfaces obtained up to now only about 0.5 extra Ba-atoms per row are
detected indicating that the disordered surface layer thickness is smaller than
about 0.3 Å, assuming a 123 composition. These nearly defect free surfaces should
be useful for device technology (21). Further from fig. 2 it is obvious that the HTSC
material is sensitive to radiation damage (22). After depositing a damage density
of 0.005 displacements per atom (dpa) the channeling increases by more than a
factor of 2 (spectra (c)). For this reason it is necessary to reduce damage by ion
bombardement especially in channneling experiments to a minimum.

   The different minimum yield values as obtained for various film / substrate
combinations may partly be attributed to the various lattice parameter misfit
values ranging between 2 and 10 % for the different combinations mentioned
above. Therefore coherency strain is expected in very thin films if commensurate
growth does occur. With increasing thickness such strain may be relieved by the
formation of misfit dislocations. Axial channeling is a valuable tool to detect strain
as well as misfit dislocations at the film-substrate interface (23). For example, in

the initial state of growth on MgO the lattice parameters of the film a (a = 3.8231 Å) will tend to expand to match the larger lattice constant of the substrate (a = 4.21 Å) and the c-axis will be compressed according to the Poisson effect. This compression will change the angle between the [013] / [103] direction in the film (accessible for channeling for c-axis oriented films) and the [110] direction in the substrate. The estimated change is several degrees and should be easily detected by axial channeling.

For films on SrTiO$_3$ with a lattice constant of 3.9015 Å the lattice misfit is much smaller (2 % for the a-direction) resulting in less strain and a corresponding small shift in the angle of about 1^0 at most. The results for a 90 Å thick film on (100) SrTiO$_3$ reveal a small shift of about 0.3^0, indicating that the layer is partially strained and that a partial commensurate growth does occur (21).

In 90 Å thick films on (100) MgO the strain has already been relieved by the formation of misfit dislocations. The higher density of misfit dislocations for films on (100) MgO is reflected indirectely in the superconducting and transport properties as well as being detectable directly by ion channeling. Metallic behaviour and zero resistance temperatures above 4.2 K were obtained in 30 Å thick films on SrTiO$_3$. Thinner films revealed temperature activated conductivity and only partial transitions to superconductivity due to inhomogeneities in the film morphology. On MgO, the critical film thickness leading to deteriorations of the transport properties was 60 Å (24).

The substrate temperature during deposition, T$_c$, is one of the most important deposition parameters for the epitaxial growth. The dependence of the quality of thin EuBa$_2$Cu$_3$O$_x$ films on T$_s$ has been studied here in detail. In previous X-ray measurements of YBa$_2$Cu$_3$O$_x$ films on SrTiO$_3$ (25) it was found that the crystalline direction of the films changes from c-axis growth to a-axis growth when the substrate temperature was reduced. As an illustrative demonstration of this growth behaviour in fig. 3 x-ray spectra are shown for three samples representing the different growth regimes.

In the following experiments the applied temperature range for EuBaCu$_3$O$_x$ on LaAlO$_3$ can be devided into three regions: first, from 650^{0}C to 740^{0}C for a-axis, second, from 740^{0}C to 800^{0}C for a / c- axes mixed, and third, from 800^{0}C to 850^{0}C for c-axis growth. The minimum yield values as a function of the substrate temperature reveals two minima of 4 % and 5 % at T$_s$ of 710^{0}C and 810^{0}C, respectively (see fig. 4). These minima were attributed to pure a- and c-axis growth as confirmed by x-ray measurements. An enhanced dechanneling yield at the interface was noted for a-axis oriented films (19) which could be reduced by applying the template layer procedure (26).

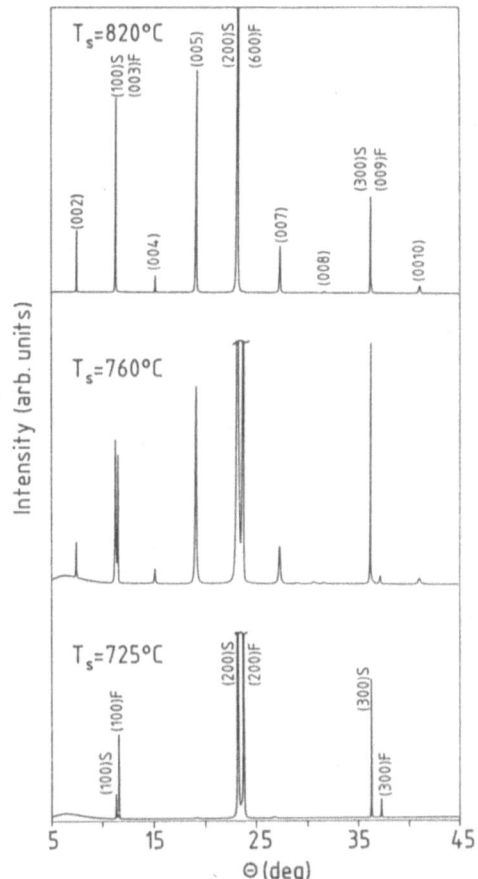

Fig. 3   Selection of X-ray spectra in the Bragg-Brentano focusing geometry
          of YBaCuO films deposited on (100) $SrTiO_3$ at different
          temperartures, T_s, showing "pure" c-axis ($T_s = 820^0C$) and a-axis
          ($T_s = 725^0C$) orientations and a mixture of both directions ($T_s = 760^0C$).

2.3 Oxygen-Sublattice

The oxygen sublattice is of major importance for the new HTSC materials,
because it is known to play a main role with respect to their superconducting
properties. Moreover the oxygen is the element with the highest mobility in this
compound and therefore may be responsible for annealing steps observed below
room temperature after low temperature ion irradiation [4]. The oxygen sublattice
is difficult to analyse, however, by RBS and channeling because of the small
scattering cross-section for oxygen in comparison to the other components and the
large background arising either from the film or the substrate. In order to enhance

91

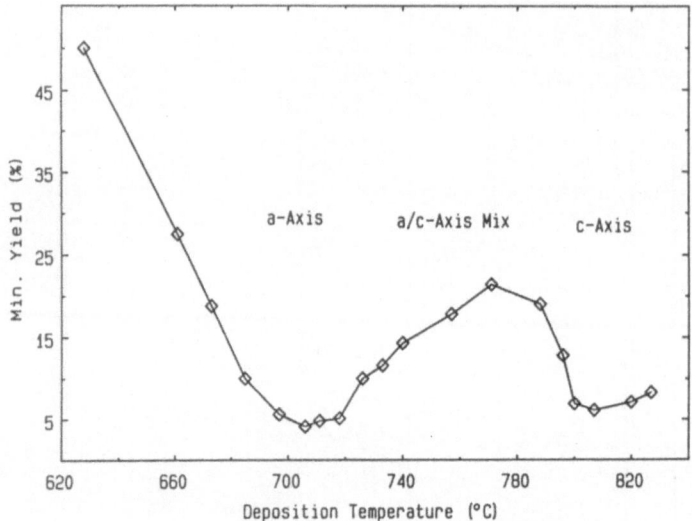

Fig. 4 χ_{min} vs. substrate temperature, T_s, for EuBa$_2$Cu$_3$O$_7$ thin films on (100) LaAlO$_3$ substrates

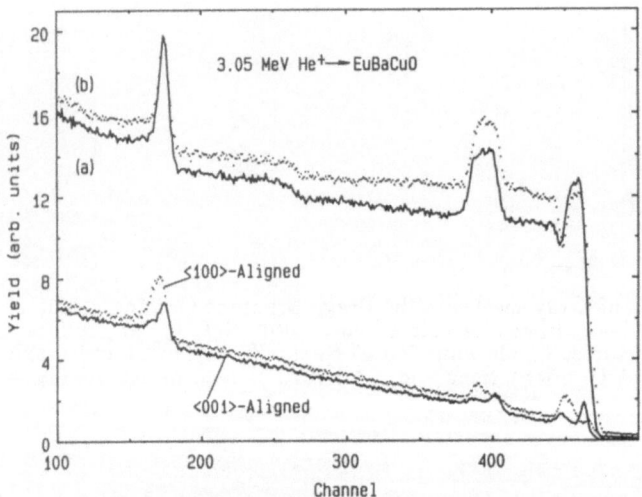

Fig. 5   Random and aligned resonant scattering spectra at 3.05 MeV He$^+$ for a-axis (dotted lines) and c-axis (solid lines) oriented EuBa$_2$Cu$_3$O$_7$ thin films on (100) LaAlO$_3$.

the detection sensitivity of oxygen the resonant cross section of He ions on ^{16}O for energies at 3.04 MeV (1) or 8.8 MeV has been used (27).

To measure the oxygen sublattice more precisely, we used the 3.04 MeV resonant scattering method. We show as an example in fig. 5 the spectra of a c- and an a-axis oriented EuBa$_2$Cu$_3$O$_7$ thin film. The minimum yield values for the oxygen sublattice are 26 % for the c-axis and 57 % for the a-axis oriented film,

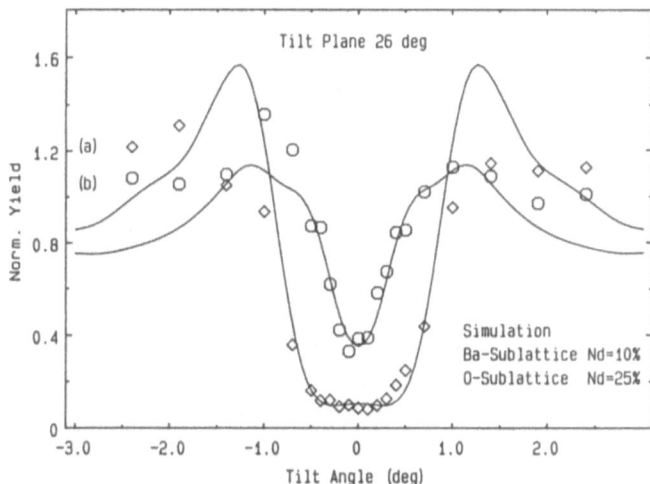

Fig. 6   Angular yield curves of (a) the Eu / Ba - and (b) the O-sublattice along a
tilt plane 26 deg. off the (100)-plane. Solid lines are from Monte Carlo
simulation calculations.

whereas for the Eu / Ba (Cu) sublattices χ_{min} values of 6.5 % (11 %) and 13 % (15 %)
are measured, respectively. Minimum yield values around 60 % in the O-sublattice
are typical for a-axis and around 25-40 % for c-axis growth. The minimum yield
values for the c-axis film in fig. 5 correspond to a random fraction N_d of about 5 %
for the Eu / Ba-, 10 % for the Cu- and 24 % for the O-sublattice. A Monte Carlo
simulation of angular yield curves in a defective structure is given together with
measured data in fig. 6. It shows a minimum yield value in the oxygen sublattice of
about 30 %. Monte Carlo simulation of a perfect lattice at 30 K, however, yield a
χ_{min} value for oxygen of 6 % which is in good agreement with measurements of the
oxygen sublattice in a bulk single crystal. The discrepancy in the dechanneling for
films and single crystal indicates considerable disturbance in the oxygen
sublattice of the films. The higher χ_{min} in a-axis oriented films may be due to the
lower deposition temperature. A possible defect structure could be connected to
additional distortions of the CuO-planes in the films, in accordance with the
increased dechanneling in the Cu-sublattice as compared to the Eu / Ba-sublattice.

In conclusion, it was shown that ion channeling and backscattering
spectrometry has provided valuable information on many aspects of the synthesis
and analysis of high quality superconducting $YBa_2Cu_3O_7$ thin films. Control and
optimization of various deposition processes and substrate surfaces resulted in a
reproducible production of high quality thin films with low disorder densities at
the substrate / film interface, within the films and at the film surface. These
epitaxial films on various substrates certainly fulfill all the requirements imposed
by special needs in science and technology.

3. Radiation response and defect properties

Ion irradiation is used to change the transport properties of superconductors in a controllable fashion. This is of interest for practical reasons as well as for the supply of basic information on the microscopic correlation between defects and superconductivity.

The high-temperature superconductors have open, layered structures which cause their physical properties to be highly anisotropic. Structural transformations and metal-insulator transitions can easily be evoked by doping in this class of material. For the well-known $YBa_2Cu_3O_{7-\delta}$ superconductor for example, the oxygen vacancy concentration governs such transitions and controls the superconductivity to a large extent. Having these facts in mind, it is not astonishing that ion irradiation resulted in a rich variety of interesting effects, which differ from those known from irradiation studies of conventional superconductors. Among these effects is the exponential increase of the resistivity, ρ, with ion fluence, ϕ, leading to a metal-to-insulator transformation. The occurrence of large recovery stages at temperatures above about 150 K of the radiation induced changes of ρ and the transition temperature, T_c, were observed in low-temperature irradiation experiments. Important structural changes occur such as the increase of the lattice parameters with ϕ, the radiation induced orthorhombic to tetragonal phase transition, and amorphization. Beneficial effects of irradiation such as enhanced pinning and patterning have also been applied. More details have been presented in recent reviews and references therein (7, 28, 29).

Thin single crystalline REBaCuO (RE = Y, Gd, Eu) films with the c-axis or the a-axis oriented perpendicular to the (100) surface of single crystal $SrTiO_3$-substrates were produced by dc-sputtering in an inverted cylindrical magnetron. Details of the preparation procedure, the analysis and the properties of the films are described in (3). Irradiations have been performed with 300 keV He^+ and 600 keV Ar^{++}-ions at RT at 15 degrees off axis. The particles penetrated the films and came to rest in the substrates. RBS and channeling studies with 2 MeV He^+ ions have been performed in situ for low-temperature irradiation and analysis, while for studies of the oxygen sublattice using the 3.05 MeV He^+ resonance, the sample had to be transferred to a separate scattering chamber. X-ray diffraction in the Bragg-Brentano geometry and rocking curves were performed to investigate the structural changes, using a rotating anode X-ray source operated at 8 KW. Line positions, their intensities and widths were analyzed for determining the lattice parameters, strain, static displacements and amorphized fractions of the films.

3.1 Transport properties

Irradiation experiments of HTSC thin films soon revealed that the T_c depression rate, the width of the resistively measured transitions, δT_c, the increase of ρ and the decrease of T_c were related to the intrinsic chemical and structural disorder of grains of different crystalline quality. Percolation effects between superconducting grains, and an enhanced radiation sensitivity of grain boundaries [31] affected the irradiation results in such a way that the T_c suppression rate and δT_c increased with decreasing oxygen content [32]. This is in contrast to results of similar studies of conventional superconductors with A15 and B1 crystal structure. For these superconductors the T_c depression rate decreased with the deviation from stoichiometric composition and grain boundaries did not affect the results to a large extent.

The main features of low fluence irradiation experiments of YBaCuO films are shown in fig. 7. Here a series of resistance versus temperature curves are presented after irradiation of a high quality single crystalline film with various fluences of 600 keV Ar^{++} ions. For low fluences the parallel shift of the transition curves and the nearly unaffected slope $d\rho / dT$, are indicative of high quality film

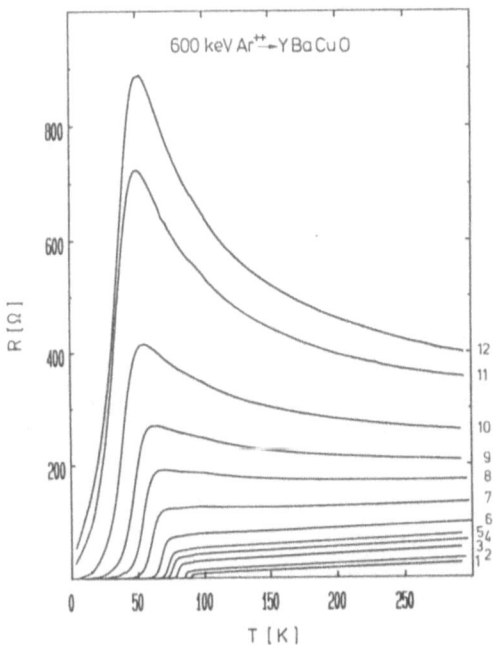

Fig. 7   Resistance vs. temperature curves of a single crystalline YBaCuO thin film irradiated with 600 keV Ar^{++} ions at RT with following fluences: 1: before irradiation; 2-12: 3, 7, 11, 14, 18, 22, 25, 28, 32, 35 and 41 $\times$ 10^{12} Ar^{++}/cm^2.

growth. In the metal-to-insulator transition regime the transition width, δT_c, broadens rapidly and the activated conduction appears. In this regime the resistance varies like $\exp(T^{-1/4})$ beween about 50 and 300 K. This relation describes the R vs. T dependence also at higher fluences, where the superconductivity is completely suppressed (not shown).

The increase of ρ with decreasing T in the temperature activated regime can be described by $\rho(T) = a \exp(Q/T^a)$, where a is found to range between 0.25 and 1 in various experiments using different irradiating particles and material of different crystalline quality. The value of a might indicate the conductivity mechanism ranging from activation across a gap ($a=1$) to localization and variable range hopping ($a=0.25$). The value $a=1$ and a conductivity maximum at about 100 K between metallic and insulating behaviour was observed after H^+-irradiation of thin single crystalline films (28, 29, 33). The R vs. T dependence observed after high dose As^+ and O^+ irradiation of textured films could also be described by $a=1$ (34). Ne^+ ion irradiation of single crystalline films resulted a value of $a=0.5$ and a Hall coefficient that varied very little (35). From these latter results it was argued that the transition is caused by localization and not by a reduction of the carrier density. A similar conclusion was drawn from results on neutron irradiated ceramic and single crystalline YBaCuO bulk samples, where an a-value of 0.25 was observed, fitting the $\rho(T)$ data over a wide range of temperatures (36).

The T_c-depression rates and the transition width are quite different for proton and Ar^{++} ion irradiated thin films. This behaviour is attributed to the different density distribution of the energy deposited during H^+ and Ar^{++} ion irradiations. The temperature dependence of the resistivity in the metal-insulator transition regime is also quite different in both cases. For the Ar^{++}-irradiated thin films ρ is

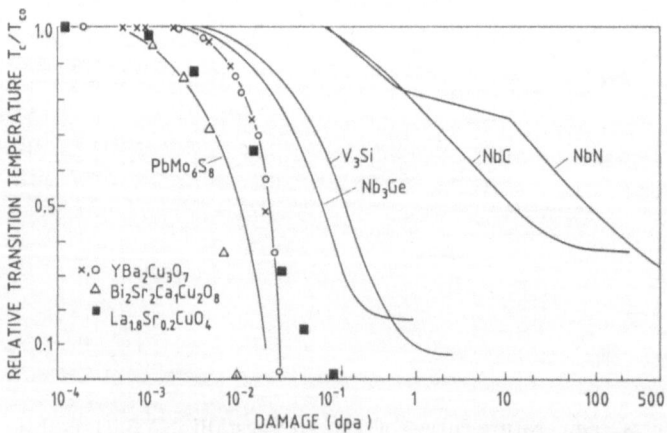

Fig. 8   Normalized superconducting transition temperature vs. damage for LaSrCuO, YBaCuO and BiSrCaCuO films in comparison to other classes of superconductors.

proportional to $\exp(T^{-1/4})$ while for proton irradiated samples ρ varies like $\exp(T^{-1})$. These results imply that the conductivity mechanisms near the metal-insulator transition depend on details of the defect structure. It is speculated that for proton irradiation displaced oxygen atoms play the main role while for Ar^{++} irradiated samples the overlap of insulating regions determines the conductivity mechanism (37).

A plot of the T_c-depressions versus a unified damage scale in displacements per atom (dpa) allows a comparison of the irradiation effects in the old and the new high-T_c superconductors. The data are displayed in fig. 8. They show a similar sensitivity of the high-T_c compounds alloys LaSrCuO and YBaCuO situated between the Chevrel and the A15 phase while the high-T_c compound BiSrCaCuO reveals a similar high sensitivity to radiation damage as the Chevrel phase.

3.2 Disordered Regions

The analysis of radiation induced defect structures will provide additional information, especially about the influence of temperature and cascade effects during defect formation. Ne^+ and Kr^+ ion irradiation at RT and in situ TEM analysis indicated that direct impact amorphization by the bombarding ion is possible, and with increasing dose a cellular microstructure appeared (38). For light ion irradiation and in situ TEM studies at 15 K the formation and evolution of defect clusters was observed which was assumed to be caused by a mobile anionic defect, associated with oxygen (39). Besides X-ray diffraction we used ion channeling to study defect structures and phase transitions (for a review see ref. 29). Channeling analysis of irradiated single crystalline thin films have been applied to determine the fraction of displaced atoms in the metal and oxygen sublattices and the increase of the mean vibrational amplitude probably due to small static displacements of the lattice atoms (28, 29).

The increase of disorder with deposited energy density is obvious from the increase of the dechanneling yield in the aligned spectra shown in fig. 9. The random level is reached after the energy density deposited by irradiation with 600 keV Ar^{++} ion has reached 0.18 dpa. X-ray diffraction analysis proves that the film is x-ray amorphous at this damage level. The onset of amorphization at damage levels much below 1 dpa implies the existence of long range forces driving the transformation like e.g. strain. The mechanisms of amorphization, however, are not yet fully clarified. In an early investigation on YBaCuO films (31) it was shown by TEM measurements that amorphization sets in at grain boundaries. We have found that full amorphization of YBaCuO films with proton irradiation at RT cannot be achieved, however, is possible with 300 keV He ions (40).

The increase of the minimum yield of EuBaCuO thin films with damage density deposited by irradiation with 600 keV Ar^{++} and 2 MeV He^+ ions at RT are

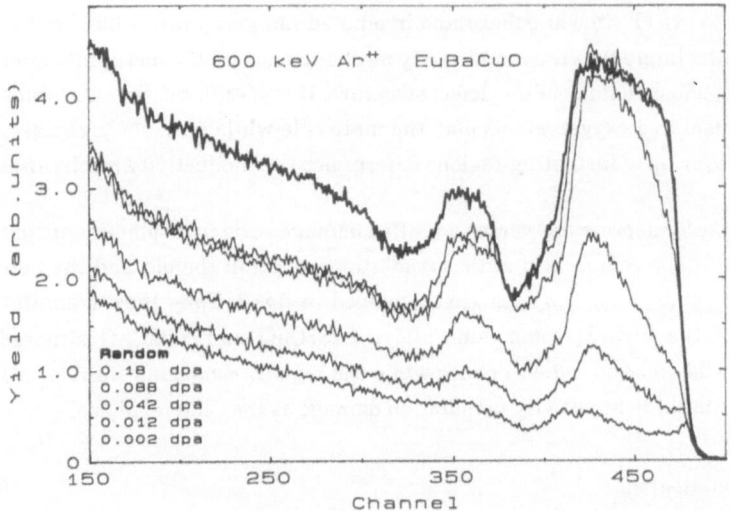

Fig. 9   Random and [001] aligned RBS spectra for an $EuBa_2Cu_3O_7$ epitaxial thin film on (100) $LaAlO_3$. The aligned spectra are shown as a function the damage density deposited by 600 keV Ar^{++} irradiation.

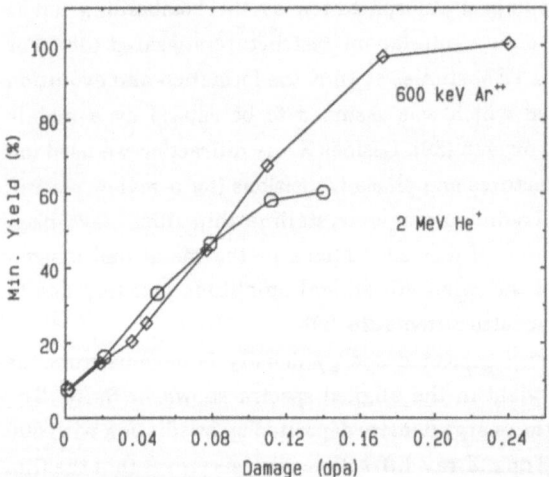

Fig. 10   Minimum yield of the Ba/Eu sublattice as a function of the damage density deposited by 600 keV Ar^{++} and 2 MeV He^+ irradiation.

shown in fig. 10. It is clearly seen that full amorphization is reached by Ar^{++} irradiation while with 2 MeV He^+ ions the random level will not be reached at low fluences. These data and those for proton irradiation imply that equilibrium is reached between defect production and annealing and that a certain defect density or cascade effects are necessary to achieve amorphization. Direct impact amorphization along the bombarding ion tracks possibly takes place by heavy ion bombardment. Channeling experiments further revealed that equal amounts of atoms are displaced on the different heavy atom sites in single crystalline YBaCuO films with He bombardment (41). It appears that further detailed investigations are necessary to clarify whether also the already existing data may be combined into a standardized model of amorphization.

3.3 Small Static Displacements

The relationship of χ_{min} and the critical angle, $\Psi_{1/2}$, with the mean vibrational amplitude $u_\perp$ is applied to determine changes of the lattice vibrations and small static displacements of the atoms from their equilibrium sites. This technique was used to determine atomic relaxations in superconducting non-stoichiometric refractory compounds (42) and in irradiated superconductors with A15 crystal structure (43). In YBaCuO single crystals steplike changes of $\Psi_{1/2}$ were observed at the transition temperature (4).

In irradiated YBaCuO thin films a narrowing of the critical angle with increasing deposited energy density was noted (28, 29) which is due to an increase of the mean vibrational amplitude including static and dynamic contributions. Similar experiments have now been performed using YBaCuO single crystals, irradiated with 1.5 MeV He$^+$ ions at RT. The results are shown in fig. 11 together with our previous results obtained for thin films. A steep decrease of the critical angle with increasing damage is noted. This is in agreement with data published recently (44), stating that the decrease of the Debye temperature is correlated with

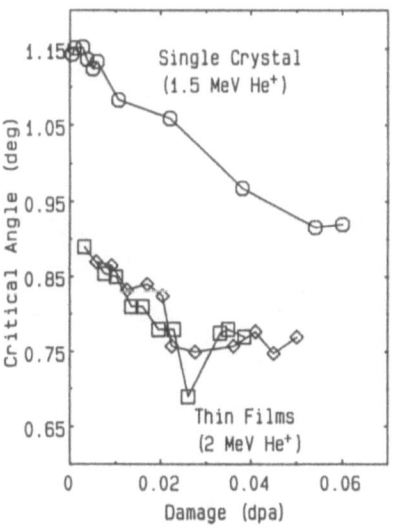

Fig. 11   Critical angle of the Ba sublattice vs. deposited energy density of YBaCuO single crystals and thin films.

a loss of elasticity. The question arose if the decrease of the critical angle is correlated with the T_c suppression which occurs in the same damage regime. We therefore studied two different films with transition temperatures of 50 K and 90 K. The decrease of the critical angle with damage is rather similar (see fig. 11). Thus it is concluded that the displacements are a general structural property of this class of material.

The enhanced vibrational amplitude contains static and dynamic contributions. In order to separate these contributions, the temperature-dependence of the critical angle has to be studied. Such measurements have been performed by irradiating and analyzing an YBaCuO single crystal with a 1.5 MeV He$^+$ ion beam at different temperatures (37). The result of this analysis is compared with Monte-Carlo simulation calculations. It could be shown that the main contribution due to irradiation is of static nature.

It should be noted that both defect structures, the amorphous fractions as well as the radiation enhanced vibrational amplitudes have been observed by x-ray diffraction using the modified Wilson Plot for analysis (29, 30). As an example the effects are demonstrated in fig. 12 where the line intensity of the (005) reflex is shown as function of the fluence of 300 keV He$^+$ ions. The line intensity decreases strongly with fluence, the line width broadens and at fluences of about 5×10^{16} He$^+$/cm^2 the line intensity is completely smeared out. The amorphization can be judged by the fact that intensity appears at an scattering angle of about 14.5 deg (not shown) in agreement with the peak position of an amorphous deposit of the

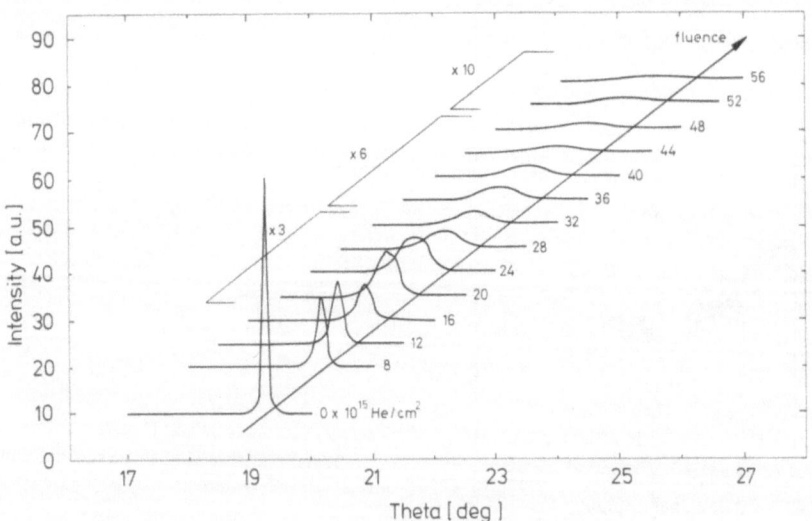

Fig. 12   Section of the x-ray spectrum showing the (005)-peak of the YBa$_2$Cu$_3$O$_7$ phase as a function of the He$^+$ ion fluence.

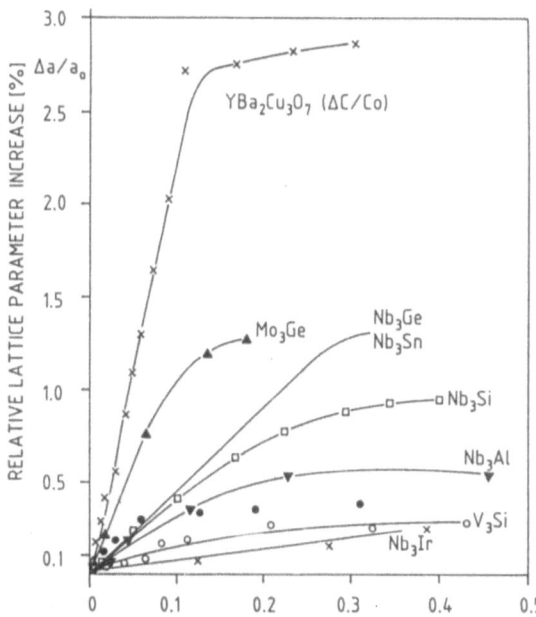

Fig. 13   Relative c-axis lattice parameter increase as a function of the deposited
damage energy in comparison to other classes of supercondutors.

YBaCuO phase (40). Further is seen that the x-ray line shifts to lower Θ-values
with increasing He ion fluence. From this shift a continous expansion of the unit
cell can be inferred. A saturation value of about 2.8 % is reached with increasing
fluence. The relative increase of the c-axis lattice parameter is shown in fig. 13
together with the swelling behavior of A15 phase material. The large increase of
the unit cell is due the observed orthorhombic to tetragonal transition and due to
defect accumulation and the local volume change that occurs during
amorphization.

3.4 Conclusions

Irradiation experiments of high-T_c oxide superconductors provided a wealth of
results which are quite unexpected if compared with those obtained for irradiated
conventional superconductors. The main reason for this behavior is attributed to
the oxygen atoms which are rather mobile and, if displaced during irradiation, can
strongly change the structural as well as the electronic properties. A transition
from the metallic-orthorhombic to a semiconducting phase can easily be enforced
by irradiation. Both, the T_c-suppression rate as well as the resistance increase rate
depend on the oxygen concentration and increase with increasing oxygen vacancy
content. There are great difficulties in a detailed physical understanding of the
irradiation results. This is partly due to our lack of knowledge on details of the

electronic structure of these complicated materials. This fact is further complicated by the existence of radiation induced phase mixtures consisting of a distorted orthorhombic phase and an amorphous phase. The latter seems to form already within the collision cascade of a single particle after heavy mass ion irradiation. For light ion irradiation the T_c reduction at low fluences occurs without much broadening of the resistively measured transition curves. This indicates that the starting material is rather homogenous as far as the chemical and structural disorder is concerned and that a homogeneous distribution of point defects prevails. It is concluded that the phase with a reduced T_c-value is a radiation induced oxygen deficient orthorhombic phase. At high fluences the temperature dependent resistance curves indicate a mixture of superconducting and semiconducting behavior, which can still be attributed to the distorted orthorhombic pase. The amorphous phase regions are insulating and shorted. The different R vs. T dependences observed for H^+ and Ar^{++} ion irradiated thin films imply different activated conductivity mechanisms. The homogeneous displacement of oxygen atoms by protons could lead to a gap at the Fermi energy while the overlap of insulating regions formed by Ar^{++} irradiation could lead to a percolative transition. For the physical understanding of the property changes more information is necessary especially on the electronic structure of HTSC prior to and after irradiation.

Acknowledgements

The author would like to thank his colleagues J. Geerk, T. Kröner, G. Linker, F. Ratzel, J. Remmel, R. Smithey and B. Strehlau who contributed significantly to the results presented in this review for a very pleasant and fruitful cooperation.

References

[1]   W.K. Chu, J.W. Mayer and M.-A. Nicolet, *Backscattering Spectrometry* (Academic Press, New York, 1987) and references therein.

[2]   L.C. Feldman, J.W. Mayer and S.T. Picraux, *Material Analysis by Ion Channeling* (Academic Press, New York, 1982) and references therein.

[3]   J. Geerk, G. Linker and O. Meyer, *Material Science Reports 4* (1989) 193

[4]   R.P. Sharma, L.E. Rehn, P.M. Baldo and I.Z. Lin, Phys. Rev. Lett. 62 (1989) 2869

[5]   R.P. Sharma, L.E. Rehn, P.M. Baldo, U. Welp and Y. Fang, Phys. Rev. B (1991) in press

[6]   L.E. Rhen, R.P. Sharma, P.M. Baldo, Y.C. Chang and P.Z. Jiang, Phys. Rev. B42 (1990) 4175

[7]   O. Meyer, *In Studies of High Temperature Superconductors.* Vol. 1, A.V. Narlikar, ed. (Nova Science Publishers, N. Y. 1989) 139

[8]   O. Meyer, B. Egner, G.C. Xiong, X.X. Xi, G. Linker and J. Geerk, Nucl.
       Instr. and Meth. B39 (1989) 628

[9]   B. Roas, B. Hensel, G. Saemann-Ischenko and L. Schultz, Appl. Phys. Lett.
       54 (1989) 1051

[10]   L. Civale, A.D. Marwick, M.W. Mc Elfresh, T.K. Worthington, A.P.
       Malozemoff, F.H. Holtzberg, J.R. Thompson and M.A. Kirk, Phys. Rev. Lett.
       65 (1990) 1164 and L. Civale et al. Phys. Rev. Lett. (1991) in press

[11]   A.E. White, K.T. Short, R.C. Dynes, A.F.J. Levi, M. Anzlowar, K.W.
       Baldwin, P.A. Polakos, T.A. Fulton and L.N. Dunkelberger, Appl. Phys.
       Lett. 53 (1989) 1010

[12]   G.C. Xiong, H.C. Li, G. Linker and O. Meyer, Phys. Rev. B38 (1989) 240

[13]   O. Meyer, J. Geerk, T. Kröner, Q. Li, G. Linker, B. Strehlau, X.X. Xi, Mat.
       Res. Soc. Symp. Proc. Vol. 157 (1990) 493

[14]   A.D. Marwick, G.J. Clark, D.S. Yee, R.B. Laibowitz G. Codeman and J.J.
       Cuomo, Phys. Rev. B39 (1989) 9061

[15]   G.J. Clark, A.D. Marwick, R.H. Koch and R.B. Laibowitz, Appl. Phys. Lett.
       51 (1987) 139 and 1462

[16]   R.H. Koch, C.P. Umbach, G.J. Clark, P. Chaudhari and R.B. Laibowitz.
       Appl. Phys. Lett. 51 (1987) 200

[17]   Q. Li, F. Weschenfelder, O. Meyer, X.X. Xi, G. Linker, J. Geerk, J. of the
       Less-Common Metals, 151 (1989) 295

[18]   F. Wang, M. Siegel, R. Smithey, J. Geerk, G. Linker and O. Meyer, E-MRS
       Straßburg, May 1991

[19]   J. Remmel, J. Geerk, G. Linker, O. Meyer, R.L. Wang. Th. Wolf, IBA 10,
       Juni 1991, to be published NIMB (1992)

[20]   N.G. Stoffel, P.A. Morris, W.A. Bonner and R.J. Wilkens, Phys. Rev. B37
       (1989) 2297

[21]   O. Meyer, J. Geerk, Q. Li, G. Linker and X.X. Xi, Nucl. Instr. and Meth. B45
       (1990) 483

[22]   O. Meyer, F. Weschenfelder, J. Geerk, H.C. Li and C.C. Xiong, Phys. Rev.
       B37 (1988) 9757

[23]   S.T. Picraux. L.R. Dawson, J. I. Tsuo, B.L. Doyle and S.R. Lee, Nucl. Instr.
       and Meth. B33 (1988) 891

[24]   X.X. Xi, J. Geerk, G. Linker, Q. Li and O. Meyer, Appl. Phys. Lett. 54 (1989)
       2367

[25]   G. Linker, X.X. Xi, O. Meyer, Q. Li and J. Geerk, Solid State Comm. 69
       (1989) 249

[26]   G. Linker et al. accepted in Physica C

[27]   J.A. Martin, M. Nastasi, J.R. Tesmer and C.J. Maggiore, Appl. Phys. Lett.
       52 (1988) 2177

[28]   L.E. Rehn, IBA 10, Eindhoven, July 1991, Nucl. Instr. and Meth. B (1992)

[29]   O. Meyer, J. Geerk, T. Kröner, Q. Li, G. Linker, B. Strehlau, X.X. Xi, Mat.
       Res. Soc. Symp. Proc. Vol. 157 (1990) 493

[30]   O. Meyer, B. Egner, G.C. Xiong, X.X. Xi, G. Linker and J. Geerk, Nucl.
       Instr. and Meth. B39 (1989) 628

[31]   G.J. Clark, A.D. Marwick, R.H. Koch and R.B. Laibowitz, Appl. Phys. Lett.
       51 (1987) 139 and 1462

[32]   D. Bourgault, D. Groult, S. Bouffard, J. Provost, F. Studer, N. Nguyen, B.
       Raveau, M. Toulemoude, Phys. Rev. B39 (1989) 6549

[33]   G.C. Xiong, H.C. Li, G. Linker and O. Meyer, Phys. Rev. B38 (1988) 910

[34]   A.D. Marwick and G.J. Clark, Nucl. Instr. and Meth. B37/38 (1989) 910

[35]   J.M. Valles, A.E. White, K.T. Short, R.C. Dynes, J.P. Garno, A.F.J. Levi, M. Anzlowar, and K. Baldwin, Phys. Rev. B39 (1989) 11599

[36]   B.N. Goschchitskii, S.A. Davydov, A.E. Karkin, A.V. Mirmelstein, M.V. Sadovskii, V.I. Voronin, Physica C 162-164 (1989) 1019, see also pages 997 and 1023

[37]   O. Meyer, T. Kröner, J. Remmel, J. Geerk, G. Linker, B. Strehlau and Th. Wolf, REI 10, Weimar Juli 91, Nucl. Instr. and Meth. B (1992)

[38]   M.A. Kirk, M.C. Frishherz, J.Z. Liu, L.L. Funk, L.J. Thompson, E.A. Ryan, S.T. Ockers and H.W. Weber, Physica C 162-164 (1989) 532

[39]   M.-O. Ruault, H. Bernas and M. Gasgnier, Philosophical Magazine B60 (1989) 801

[40]   G. Linker, J. Geerk, T. Kröner, O. Meyer, J. Remmel, R. Smithey, B. Strehlau, X.X. Xi, Nucl. Instr. and Meth. B (1991) accepted

[41]   O. Meyer, F. Weschenfelder, X.X. Xi, G.C. Xiong, G. Linker and J. Geerk, Nucl. Instr. and Meth. B35 (1988) 292

[42]   R. Kaufmann and O. Meyer, Phys. Rev. B28 82983) 6216

[43]   O. Meyer, R. Kaufmann, B.R. Appleton, and M. Chang, Solid State Comm. 39 (1981) 825 and references therein.

[44]   L.E. Rehn, R.P. Sharma, P.M. Baldo, and Y. Fang, Nucl. Instr. and Meth. B (1991) accepted

Slow Particle-Induced Electron Emission from Above and Below Metal Surfaces

H. Winter

Institut für Allgemeine Physik, Technische Universität Wien,
Wiedner Hauptstraße 8–10, A-1040 Wien, Austria

Electron emission caused from bombardment of metal surfaces with slow heavy particles can be initiated by the kinetic as well as potential energy of the projectiles.

After shortly recapitulating our present knowledge on kinetic ("KE") and potential emission ("PE") processes, we describe a new experimental technique for their study by virtue of the related electron emission statistics ("ES").

The value of ES measurements will then be demonstrated in several case studies on the following topics.

(a) Precise determination of threshold impact energy for kinetic emission.

(b) Comparison of kinetic electron emission induced by neutral vs. ionized atoms and by atomic vs. molecular ions ("projectile shielding effects").

(c) Evaluation of exclusive potential electron emission properties for impact of multicharged projectiles above the KE threshold.

1. Slow particle-induced electron emission - basic concepts

If the surface of a solid, in particular of a clean or gas-covered metal, is bombarded by slow neutral or ionized atoms/molecules, electrons can be emitted due to transfer of potential energy (*potential emission - PE*) and/or kinetic energy (*kinetic emission - KE*) from the projectiles onto the target atoms and electrons. Such slow particle-induced electron emission ("sPIE") processes are of both fundamental practical interest, and have therefore been investigated to considerable depth (see /1/ and /2/ for reviews on PE and KE, respectively).

Both classes of processes depend very sensibly on the metal surface conditions. In first approximation, PE can be related to electronic transitions between projectile and surface before the impact has taken place, whereas KE is initiated only after the particle has made a close contact with the surface. In a somewhat more detailed view, PE arises from Auger-type processes which take place outside ("above") the surface during time intervals of the order of typically 10^{-14} s, which are comparable to the flight times of relatively slow ($v \leq 10^5$ m/s) particles within the distance where the first electronic transitions between the projectiles and the metal surface can start, and the surface impact takes place (typically several 10^{-10} m for singly charged ions, cf. /1/). However, for higher projectile velocity ($v \geq 1$ a.u.) there remains not enough time for a complete neutralization and deexcitation of the projectile until its surface impact. PE is therefore most efficient at the lowest impact energy and involves no impact energy threshold. The KE process, on the other hand, is related to the stopping power of the projectile within the uppermost atomic layers of the solid, from where electrons can escape which have been set free inside the solid(i.e. "below" the surface). The KE process is subject to a projectile impact energy threshold.

From these simple considerations we see that for the projectile energy range of greatest practical interest (0,1 - 10 keV/amu, corresponding to $0,06 \leq v \leq 0,6$ a.u., with 1 a.u. $\approx 2 \times 10^6$ m/s), the

PE and KE processes should commonly become interrelated, which severely limits the physical relevance of our above presented simple concepts.

Recently, we have achieved to measure the statistics for particle-induced electron emission ("ES"), i.e. to determine the probabilities W_n for emission of a given number n (n = 1, 2, etc.) of emitted electrons. As compared to the usually measured total electron emission yields and/or energy distributions of emitted electrons /1, 2/, these ES can provide a considerably clearer distinction among different processes which are contributing to the particle-induced electron emission. This involves not only the separability of processes initiated by respectively potential and kinetic projectile energy transfer, but also permits a number of other interesting investigations in the field. For instance, the influence of differently charged projectiles (neutral vs. ionized vs. multiply charged) or of atomic vs. molecular projectiles on the sPIE process can now be studied in considerably greater detail. Furthermore, rather precise measurements of very small total electron emission yields, as pertinent to the KE threshold region, have now become possible. Results of these and related investigations already have provided an improved understanding of the various processes contributing to sPIE, as we will demonstrate in the following chapters.

2. Measurement of slow particle-induced electron emission statistics

PIE from solid surfaces is usually studied by considering the related total electron yield γ and/or ejected electron energy distribution $K(E_e)$. These properties are of great practical interest but do not directly give access to a more detailed understanding of the mechanisms responsible for electron emission. The emission statistics is related to the total emission yield by

$$\gamma = \sum_{n=0}^{\infty} n \cdot W_n \; ; \quad \sum_{n=0}^{\infty} W_n = 1 \qquad (1)$$

We have described in detail elsewhere how the emission statistics can correctly be measured and evaluated for both ionized and neutral projectiles /3/. Fig. 1 shows our most recently developed setup /4/ which is particularly suited for low ion impact energies (E ≥ 50 q eV). The ions can be

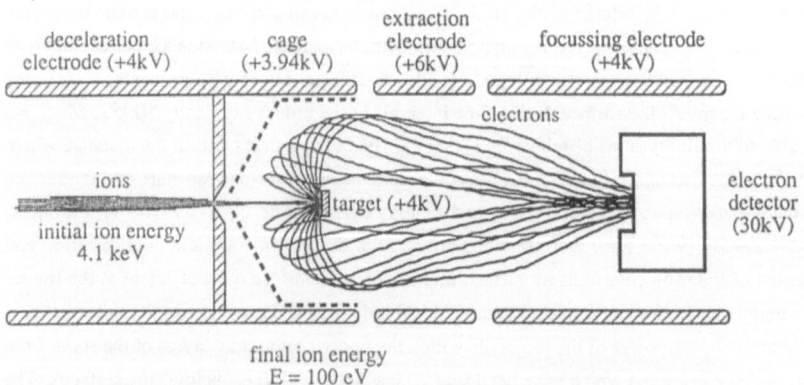

Fig. 1: Experimental setup for measuring particle-induced electron emission statistics and total emission yields for impact of slow (≥ 50 eV) ions on a clean metal surface /4/.

deccelerated just in front of the target, which is situated inside a highly transparent conical electrode serving for deflection of the emitted electrons toward a solid state detector situated behind the target. The electrons are accelerated into this detector with an energy of up to 25 keV by means of a three cylinder lens, with all other features the same as for our earlier developed detector system /3/. With our latest setup we have also succeeded to measure very small total electron yields ($\geq 10^{-4}$ electrons/projectile) in the following way. Independent from ion deceleration inside the detector the ion beam intensity has been precisely attenuated by a variable factor of up to 10^6 with a magnetic quadrupole lens in the ion beam line far upstream of the ES detector. With ion currents of the order of nA and for $\gamma \geq 0.1$, the total yields have been determined from ion- and electron current measurements in the usual way (cf. /3/).

With the ion fluxes then strongly attenuated to typically 10^3 particles/s, they could still be measured, but now by counting all electrons emitted from the target and using the related, already determined values of γ (see above). Since our setup - within certain limits - could assure ion deceleration without particle losses, a once selected ion flux remained fixed also if decelerated. By measuring the related electron fluxes at various ion impact energies involving the such fixed ion fluxes, γ could rather precisely be determined down to such low impact energies where measurements by standard techniques are no more feasible because of far too small electron yields.

3. Measured electron emission statistics

The statistics of particle-induced electron emission is of general interest for small particle flux measurements via counting, because the probability W_0 for emission of no electron is the related counting loss. In the relevant literature it is quite commonly assumed that the here involved counting statistics correspond to Poisson distributions P_n with their mean value γ

$$P_n(\gamma) = (\gamma^n/n!)\, e^{-\gamma} \qquad (2)$$

Since our ES measurements deliver only relative values for W_1, W_2, etc., their absolute calibration requires an independent determination of W_0, which can either be achieved by a direct measurement of γ (see above) and application of equ. (1) or by fitting the measured relative W_n values for $n \geq 1$ to a standard statistical distribution (e.g. a Poissonian) and obtaining W_0 by extrapolation. However, in accordance with studies by some other groups we could show that our measured ES generally disagree with Poisson distributions /3, 5/. This can be easily demonstrated by comparing the ratios of measured relative probabilities W_{n+1}/W_n with the corresponding ratios P_{n+1}/P_n of a Poisson distribution with mean value γ, which according to equ. (2) are given by

$$P_{n+1}/P_n = \gamma/(n+1) \qquad (3)$$

Such a comparison, which will not be affected by errors resulting from ES absolute calibration (cf. above), has been made e.g. for impact of H+ ($1 \leq E \leq 16$ keV) on clean gold, as shown in fig. 2. This particular collision system is well suited for such checks because PE contributions to the total yields are less than 0,02 (/4/, see also fig. 6a) at the involved impact energies and can thus be safely neglected in the present context. At low impact energy the measured ES are apparently narrower than the Poisson distribution, whereas the opposite is found at higher impact energy.

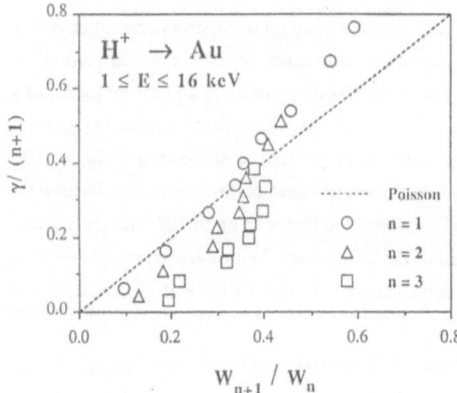

Fig. 2:
Demonstration of disagreement between measured electron emission probabilities (symbols - experimental errors smaller than symbol size) and a Poisson distribution related to the measured total electron yield.

This clear disagreement leads to the important consequence that particle counting losses evaluated by assuming Poisson-shaped ES can become rather incorrect, if the involved γ values remain not considerably larger than unity. The often made assumption of Poisson type ES is probably inspired by an assumed similarity between the processes of thermionic and particle-induced electron emission. As we have discussed in /3/, however, at low impact energy such a similarity is not present. Consequently, we always calibrated our relative ES with independently measured electron yields as described above.

The consideration of ES is especially useful if different processes contribute independently to particle-induced electron emission, since then the "apparent" ES are convolutions of the ES originating from individual mechanisms. If, by chance, an ES for one of these individual contributions can be independently determined, the not directly accessible ES for the sum of all the remaining contributions can by derived by a deconvolution procedure, as will be exemplified below for a number of cases.

4. Precise determination of threshold impact energy for kinetic emission

As already mentioned in the introduction, the course of the total electron emission yield on the impact energy is of foremost practical interest but does in general not permit a precise judgment on the relative importance of PE and KE contributions. As an example, fig. 3 shows γ data which have been determined for impact of Ne^+-, Ar^+- and Xe^+ ground state ions on clean gold and impact energies down into the region of exclusive PE /4/. The expectable PE contributions have been estimated from semiempirical calculations and indicated by horizontal dashed lines. In particular, the potential energy of Xe^+ (W ≤ 12,1 eV) is apparently too small to cause any measurable PE contribution. Fig. 3 also contains earlier measured data for Xe^+ /6/, which toward low E deviate from our results for reasons discussed in /4/. We remark that our measurements can be extended to very low γ values which so far could not be determined with comparable accuracy by any other means. As far as the KE threshold is concerned, fig. 3a (total yields) does not offer much help. However, from fig. 3b which shows ratios of relative ES probabilities for emission of n = 2 and 1 electrons, respectively, the KE threshold impact velocity can be clearly identified since PE involves the emission of one electron only, as is obvious from the respective low impact energy data.

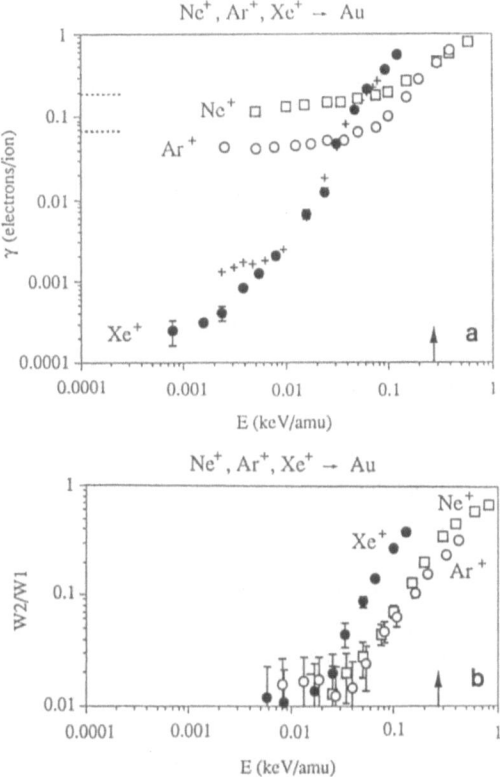

Fig. 3: Total electron yields (a) and electron emission probability ratios (b) for impact of Ne+, Ar+, Xe+ on clean gold /4/. Horizontal dashed lines indicate estimated PE contributions and vertical arrow the "conventional" KE threshold impact velocity.

Simple concepts of the KE process include the assumption that the threshold impact energy is reached where the minimum kinetic energy transfer in head-on collisions between the projectile particle and quasi-free metal electrons just overcomes the metal surface work function (see e.g. /7/). The related "conventional" KE threshold impact energies have been marked in figs. 3a,b by a vertical arrow. As clearly visible, non-negligible KE contributions are observed well below this "conventional" KE threshold, and obviously they become somewhat more important for heavier projectiles.

This low impact energy KE contribution may result either from close collisions of projectiles with individual target particles at the uppermost surface layer (electron emission from transiently formed autoionizing quasimolecules), or from so far not clearly conceived interactions of projectiles with the surface electronic states of the solid. Further studies of this kind will be useful to clarify the influence of both projectile and target species on this low impact energy KE contributions.

5. "Projectile shielding" effects in sPIE

Projectiles of given species and kinetic energy may interact in a slightly different fashion with solids if they carry a different number of electrons with them. This becomes apparent for impact of singly ionized ions in comparison with neutral projectiles, if the ions actually can reach the surface without having already undergone a resonance- or Auger neutralization processes. In fig. 4 we compare total electron yields for H^+-, H^0- and H^- impact on clean gold measured at impact energies between 1 and 16 keV /8/.

1 keV protons have already such a high velocity that within the distance of probable neutralizing transitions (some 10^{-10} m, cf. /1/) the available time of less than 10^{-14} s becomes too short to permit such a neutralization. In our experiments we found a clearly measurable difference in total yields for impact of H^+ and H^0, respectively. Furthermore, as compared to H^0 impact, for H^- the total electron yield remains apparently equal up to ca. 8 keV, but then increases faster with E, cf. fig. 4. This can be understood if we assume the splitting off of one of a loosely bound electron from H^- upon its surface impact, after which this then free electron can contribute independently to the KE yield in the same way as an equally fast but originally free electron in electron-induced ("secondary") electron emission.

The experimental results shown in fig. 4 could be well reproduced by calculations, taking into account both charge-changing of the projectiles during their penetration into the solid and the influence of these changes on the electron production inside the solid. The interaction of projectiles with the solid involves therefore a penetration depth-dependent electronic stopping power related to the quasi-continuous change of the electron shielding /9/. For a gold target the projectile equilibrium charge inside the solid is reached after passing several atomic layers. Within this region the local projectile charge state remains strongly related to the initial one, and a considerable part of the electrons being produced in this uppermost region can escape into free space.

Effects due to such an "electronic shielding" of projectiles inside the solid are also responsible for socalled "molecular effects" as commonly observed in kinetic emission /2/, and could be demonstrated by a comparison for impact of respectively atomic and molecular hydrogen ions on clean gold. Both the yields (by summation) and the ES (by convolution, cf. sect. 6) for molecular ions can only be synthesized from corresponding measurements for the equally fast atomic projectile constituents, if the correct number of accompanying electrons is taken into account. This

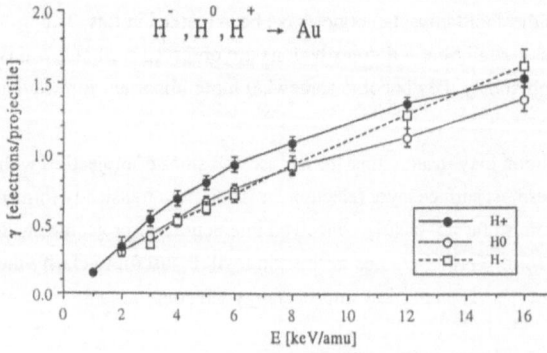

Fig. 4: Total electron yield for impact of respectively H^+, H^0 and H^- on clean gold /8/.

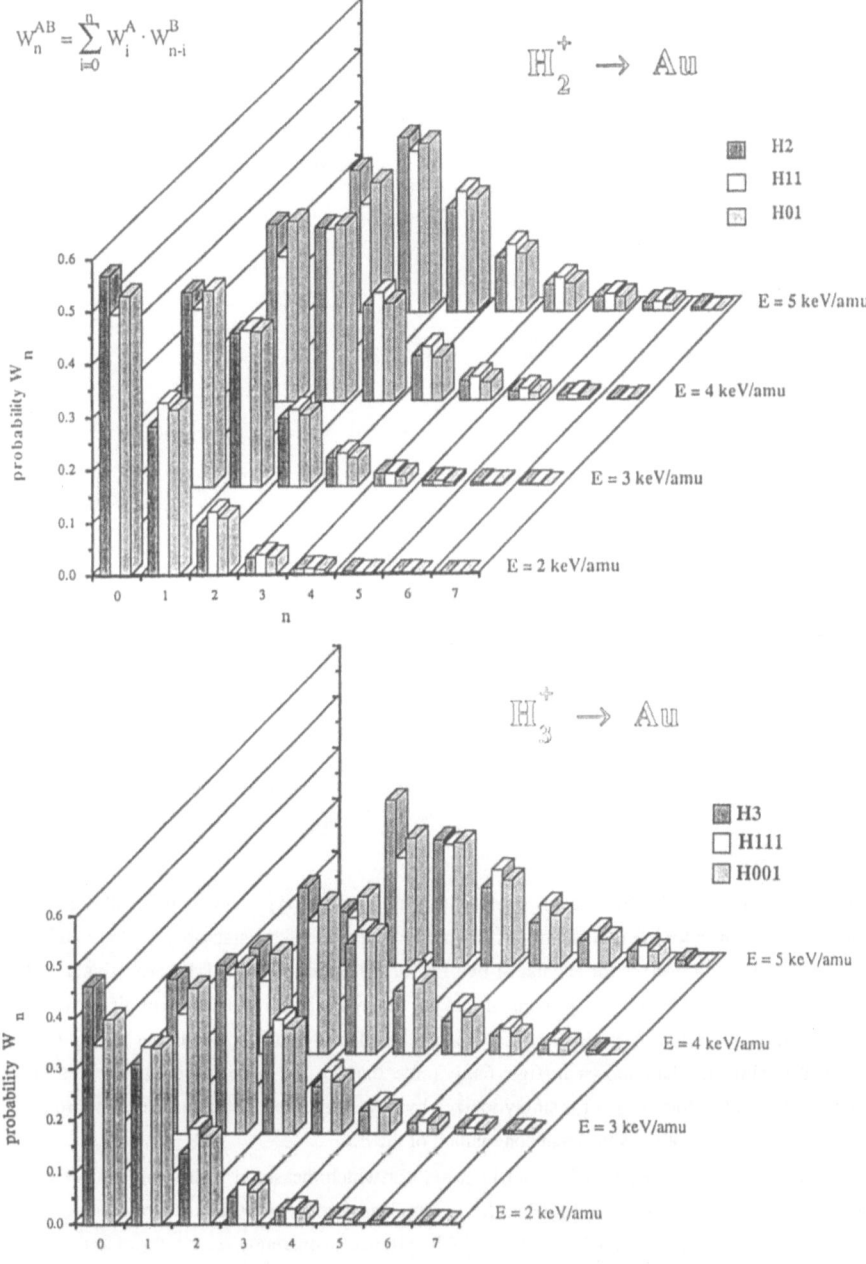

Fig. 5: Measured and synthesized electron emission statistics for impact of molecular hydrogen ions on clean gold (further explanations cf. chapter 5).

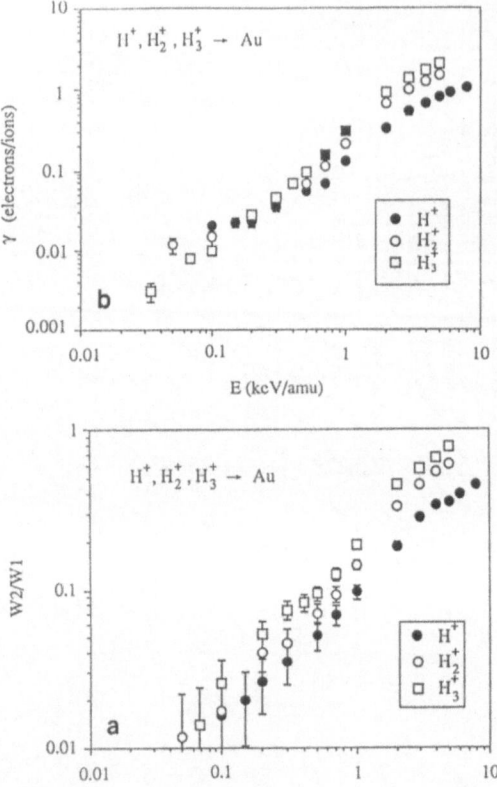

Fig. 6:
Total electron yields (a) and electron emission probability ratios (b) for impact of atomic and molecular hydrogen ions on clean gold /10/.

is shown in figs. 5a,b for ES related to impact of singly ionized H_2 - and H_3 molecules on clean gold. ES measured for the molecular ions can only be reproduced satisfactorily by convolution of one ES for H^+ and one ES for H^o and of two ES for H^o and one ES for H^+, respectively. If for such convolutions ES only for protons are taken, the sythesized results clearly differ from the measured ones. Similar investigations have been extended to considerably lower impact energy /10/. By using the data shown in figs. 6a,b (note that for all three projectile species PE is negligibly small), total yields for singly ionized hydrogen trimers could be precisely reproduced both via summation of the yields and convolution of the ES.

This is demonstrated for the hydrogen trimer ion, for which measured yields could be very well reproduced from the corresponding data for two hydrogen dimer ions and one proton, cf. fig. 7. The above presented concept of electronic shielding remains apparently correct down to rather low projectile impact energies.

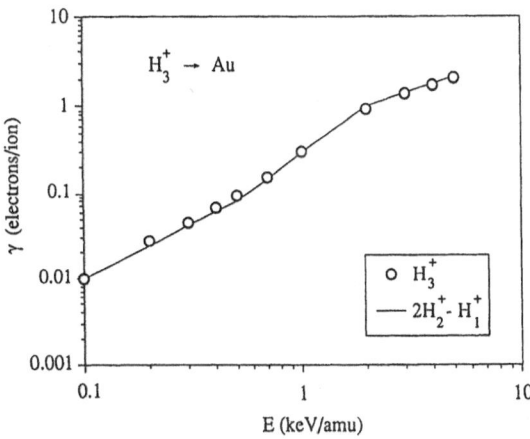

Fig. 7: Total electron yields (measured/open symbols and synthesized/cf. chapter 5) for impact of H_3 ions on clean gold vs. impact energy E /10/.

6. Electron emission induced by slow multicharged ions above the KE threshold

Fig. 8 shows raw ES data for impact of Ne^+ ions (the evaluated data have already been shown in fig. 3) on clean gold, with impact energy varied from 100 eV up to 16 keV.

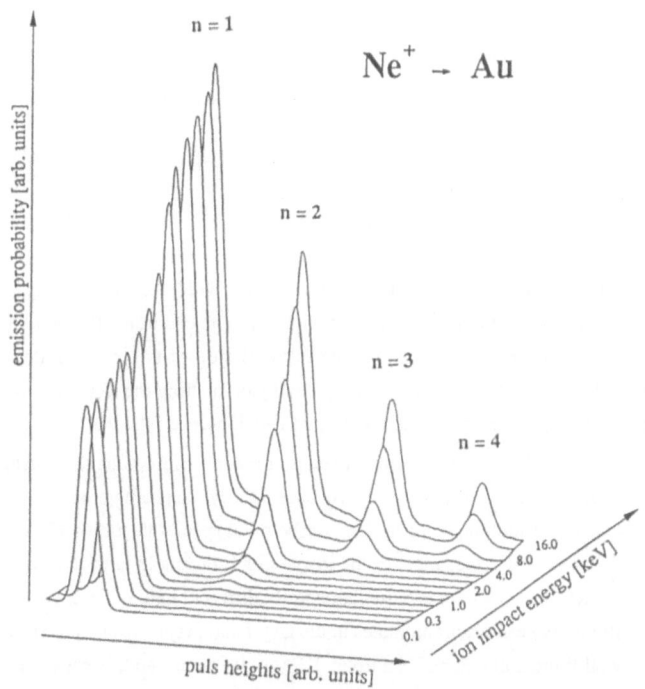

Fig. 8: Raw electron emission statistics data for Ne^+ impact on clean gold.

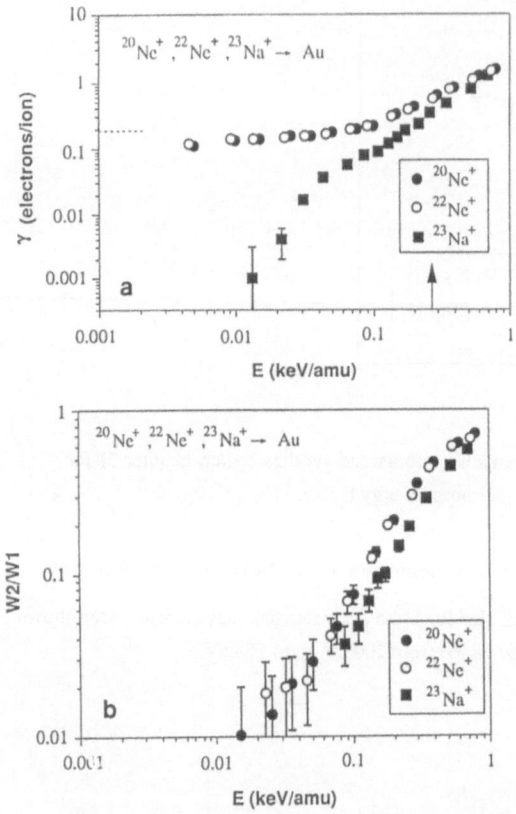

Fig. 9: Total electron yields (a) and electron emission probability ratios (b) for impact of Ne⁺ and Na⁺ on clean gold, vs. impact energy /4/.

As clearly observable, at the lowest impact energy only one electron is being ejected, but with increasing E gradually also emission of 2, 3 etc. electrons is coming up. Therefore, at low impact energy PE contributes exclusively to the total electron yield with only one electron, although the available potential energy ($W \leq 21,6$ eV) would as well permit emission of two or even three electrons. This remarkable behavior has been further investigated by comparing the total electron yields and corresponding ES for impact of equally fast Ne⁺- and Na⁺ ions, see figs. 9 a,b.

For Na⁺ impact, the involved potential energy is too small to cause PE and therefore the total yield gradually disappears toward low E, whereas for Ne⁺ γ approaches a fairly constant value toward low E which is the respective PE contribution. Nevertheless, for both ion species the ratios W_2/W_1 disappear at practically equal low E, which thus must be the involved KE threshold impact energy. Obviously, the ES for PE and KE processes are of a fundamentally different nature. PE involves relatively well defined transitions between electronic states in the target and projectile and therefore gives rise to emission of a well limited number of electrons. KE, on the other hand, is caused by dissipation of the projectile kinetic energy among a relatively large number of target electrons, the

114

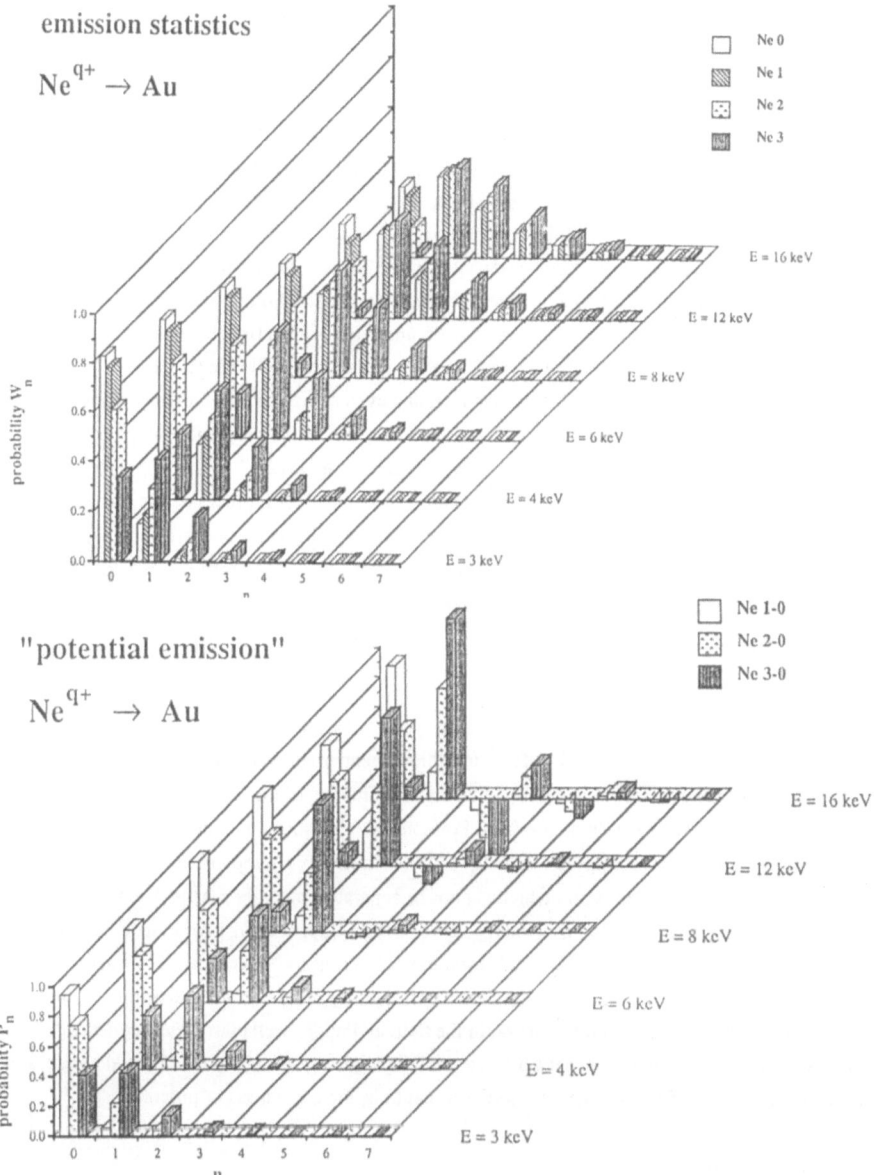

Fig. 10: Deconvolution of emission statistics for differently charged Ne ions from emission statistics for equally fast Ne atoms on clean gold (further details given in chapter 6).

number of which increases rapidly with the total transferred kinetic energy and thus with E.
Similar studies have been made with projectiles in a number of different charge states.
For example, figs. 10a,b show corrected ES for Ne^{q+} ions in charges states $1 \leq q \leq 3$, which have been derived by folding out from the actually measured ES for the various projectiles the ES

measured for neutral Ne projectiles with equal velocity (for a more detailed description cf. /11/). Fig. 10b shows that this deconvolution works only well below impact energies of ca. 8 keV, which can be understood in the following way. At low ion impact energy there remains sufficient time for a complete neutralization of the projectile until its impact on the surface. The immediately following KE is therefore equivalent to the one for an originally completely neutral particle of the same atomic species. In such a situation the PE and KE contributions will remain mutually independent and their respective ES should therefore be separable via ES deconvolution as described above. However, at higher impact energy the projectile cannot become completely neutralized anymore before its surface impact, and its KE will therefore differ from the one for equally fast neutral projectiles. Under these circumstances a deconvolution procedure can no more yield meaningful results, since the processes of PE and KE have not only become interrelated, but the KE now also depends (via impact energy) on the initial projectile charge state q.

Another interesting observation concerns the fact, that even at the lowest accessible impact energy we have not been able to unfold the measured ES for a given initial projectile charge state q from the ES measured for the same species in a lower initial charge state $q' = q - 1$, with $q' \geq 1$. This is clear evidence that the multiple resonance transitions by which the PE processes are started, rapidly neutralize the projectiles, which then via autoionization will decay under (multiple) electron emission, instead of undergoing step-by-step neutralizations down the charge state ladder from q-1 toward q = 0. More detailed discussions on such PE processes can be found in /12, 13/.

Summary

In the present notes we have demonstrated the use of the statistics for slow particle-induced electron emission from clean metal surfaces to achieve a better understanding of such processes. In particular we have explained the measurement and evaluation of the emission statistics, and have demonstrated their apparent deviation from Poisson-type distributions. Furthermore, we showed their applicability to distinguish among several mutually independent contributions to particle-induced electron emission which cannot be separated by other means. By measuring the emission statistics at very low impact energy, the threshold impact energy for kinetic emission can be precisely determined and rather small total electron emission yields can be measured more accurately than by other known methods. Consequently, ES studies constitute a valuable supplement to the methods so far common in the field and might well clarify further details of the processes of interest. We stress that ES can as well be studied for contaminated metal surfaces or insulators and with different projectile species as neutrals, ions, electrons or photons.

Acknowledgments

The author thanks Drs. F. Aumayr and G. Lakits for their important contributions to the here described studies which have been supported by Austrian Fonds zur Förderung der wissenschaftlichen Forschung (Projekt. Nr. P6381PH) and Kommission zur Koordination der Kernfusionsforschung at the Austrian Academy of Sciences.

References

1   H.D. Hagstrum, Phys.Rev. **96**(1954)325; 336
    P. Varga, Comments At.Mol.Phys. **23**(1989)111

2   J. Schou, Phys.Rev. B **22**(1980)214
    P.C. Zalm and L.J. Beckers, Philips J.Res. **39**(1984)61; Surface Sci. **152/153**(1985)135
    D. Hasselkamp, Comments At.Mol.Phys. **21**(1988)241
    W.O. Hofer, Scanning Micr. Suppl. **4**(1990)

3   G. Lakits, F. Aumayr and H. Winter, Rev.Sci.Instrum. **60**(1989)3151
    F. Aumayr, G. Lakits and H. Winter, Appl. Surface Sci. **47**(1991)139

4   G. Lakits, F. Aumayr, M. Heim and H. Winter, Phys.Rev. A **42**(1990)5780

5   G. Lakits, F. Aumayr and H. Winter, Phys. Letters A **139**(1989)395

6   E.V. Alonso, M.A. Alurralde and R.A. Baragiola, Surface Science **166**(1986)L155

7   E.V. Alonso, R.A. Baragiola, J. Ferrón, M.M. Jakas and A. Oliva-Florio,
    Phys.Rev. B **22**(1980)80

8   G. Lakits, F. Aumayr and H. Winter, Europhys. Lett. **10**(1989)679

9   G. Lakits, A. Arnau and H. Winter, Phys.Rev. B **42**(1990)15

10 M. Heim, diploma thesis, TU Wien (1990, unpublished)

11 G. Lakits and H. Winter, Nucl.Instrum.Meth.Phys.Res. B **48**(1990)597

12 H.J. Andrä, Nucl.Instrum.Meth.Phys.Res. B **43**(1989)306

13 H. Winter, Z.Phys. D: Atoms, Molecules and Clusters (1991, in print)
    P. Varga and H. Winter in Particle-induced Electron Emission,
    Springer Tracts in Modern Physics (1991, in print)

Electrons Captured and Emitted
by Highly Charged Ions near Surfaces

R. Morgenstern, L. Folkerts, and J. Das

Kernfysisch Versneller Instituut, Rijksuniversiteit Groningen,
Zernikelaan 25, 9747 AA Groningen, The Netherlands

1 Introduction

Ion beams offer a broad spectrum of properties which can be varied in order to induce different effects during interaction of these beams with solid matter. By choosing mass, kinetic energy and angle of incidence of ions impinging on a solid one can for instance vary the penetration depth, the fraction of scattered and sputtered particles or the amount of damage in a crystal.

Another important parameter is the ionic primary charge, and with the advent of sources which can produce intense beams of highly charged ions, this parameter has become subject of many investigations. An obvious advantage of using multiply charged ions is the possibility to increase the kinetic energy of the beam particles by a factor of q without having to increase the accelerator voltage. A more interesting feature however is the potential energy which is introduced into the ion-solid interaction by the high projectile charge, and one may speculate about the effect of this: does Coulomb explosion occur, when large numbers of electrons are sucked away from a small surface area by the multiply charged ion? Is it possible — by choosing the appropriate ion species and charge — to induce element selective sputtering in an alloy or to modify implantation depth profiles? In order to allow a correct choice of the parameter q of the ions one has to understand its influence on the electronic interaction between projectile and matter.

It is especially the ion-surface interaction which can be expected to be influenced by the projectile charge, since a stationary charge state distribution is reached after penetration of only a few atomic layers within the solid. In Groningen we have investigated this ion-surface interaction during the last years by studying the charge state and the intensity of scattered and sputtered particles and by analyzing the electrons resulting from the ion-surface interaction. In the following it will be shown that it is especially the information extracted from the electron energy spectra which allows one to reconstruct the history of sequentially occurring processes when the ions approach the surface.

2 Experiment

The experiments have been performed in an apparatus described earlier [1]. Multiply charged ions are extracted from an Electron Cyclotron Resonance (ECR) ion source of the MINIMAFIOS type [2] installed at the Kernfysisch Versneller Instituut (KVI) in Groningen [3]. The ion energy can be varied in a wide range by operating the ECR source on a potential variable between 2 and 20 keV and by floating the apparatus independently on a positive or negative potential up to 5 keV. The apparatus is equipped with a manipulator for the solid target and is kept at ultra high vacuum (base pressure $2 \cdot 10^{-8}$ Pa). Energy spectra of ions or electrons resulting from the ion bombardment of the target can be measured by a hemispheric electrostatic analyzer. In addition we can measure energy distributions of neutral scattered or sputtered particles by means of a time-of-flight (TOF) tube, installed at an angle of 30 deg with respect to the projectile beam. In the first place de Zwart et al [4] have measured sputtering yields of a Si target in this apparatus by collecting sputtered particles on a foil, which was subsequently analyzed by PIXE (Proton induced X-ray emission). More recently electron emission from metal surfaces has been investigated, in particular from polycrystalline tungsten surfaces.

3 Yield of sputtered particles and ejected electrons

The results of the Si sputtering measurements were surprising: the sputtering yield showed practically no dependence on the projectile charge. This is shown in fig. 1 where the amount of sputtered Si is shown for Ar projectile charges between $q=1$ and $q=9$. Apparently the electronic structure of the Si target can sufficiently fast adapt to the approaching ion such that Coulomb explosion does not occur. As opposed to this the yield of secondary electrons strongly depends on the charge of the impinging ions. This can be seen in fig. 2 where the number of ejected electrons per impinging ion is plotted as a function of the potential energy W, stored in the q-fold charged projectile, where W is taken as the sum of all ionization potentials $IP(q')$ up to $q' = q - 1$. For charge states up to $q = 8$ a linear dependence of the yield on the potential energy W is observed. For higher q values the increase of the yield becomes smaller, and it will be shown below that this deviation from the linear dependence can be ascribed to the fact that additional potential energy, brought into the collision by the higher charge states, is partly converted into increased kinetic energies of ejected electrons instead of an increased yield.

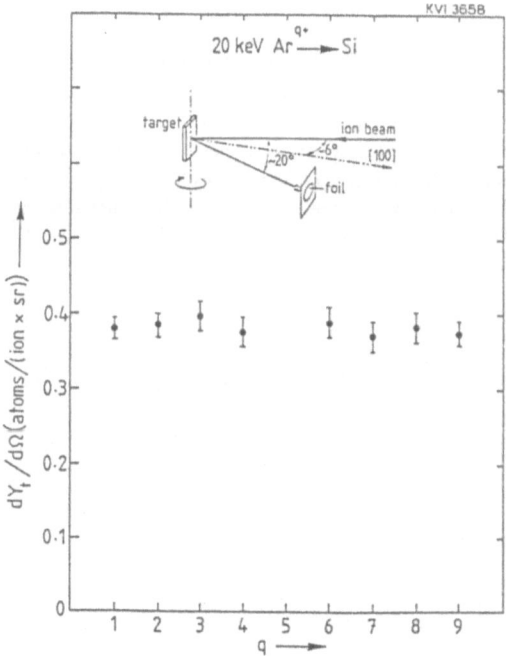

Figure 1: Yield of sputtered Si particles resulting from bombarding a Si target with Ar ions as a function of the ion charge q, from ref. [4].

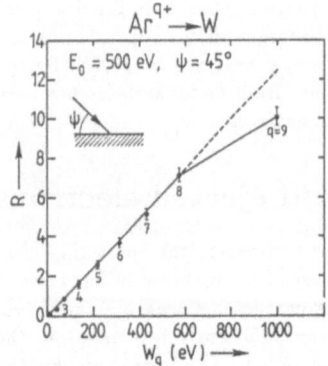

Figure 2: Yield of ejected electrons as a function of the potential energy contained in the multiply charged projectile ion.

4 Energy spectra of ejected electrons

Electrons are ejected at various stages during the collision, and their energies carry information about the electronic structure of the collision system at the moment of their ejection. As opposed to this the final charge states or energies of projectiles or sputtered particles contain only an overall information, integrated over the whole collision process. The analysis of electron energy spectra can therefore give the most detailed information on the ion surface interaction. For ions with charge states $q > 15$ also fluorescence yields and radiative transition rates become sufficiently high such that photons are ejected at different stages of the collision process. Such photons carry the detailed information as well, as has for the first time been demonstrated by Briand et al [5].

We have studied hydrogen like projectiles C^{5+}, N^{6+}, O^{7+} and Ne^{9+}. An energy spectrum of electrons arising from the impact of Ne^{9+} ions on a tungsten surface is shown in fig. 3. It consists of a broad continuous distribution of low-energy electrons and two discrete structures around 100 eV and 750 eV respectively. These discrete structures can be ascribed to LMM- and KLL-Auger processes respectively, during which the L- and K-vacancies of the projectile are filled. This implies that the electrons are ejected from

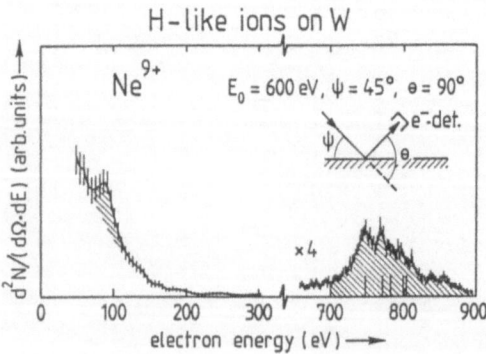

Figure 3: Energy spectrum of electrons arising from Ne^{9+} collisions on a tungsten surface at a kinetic energy of 600 eV and an angle of incidence of $\Psi = 45\,\mathrm{deg}$.

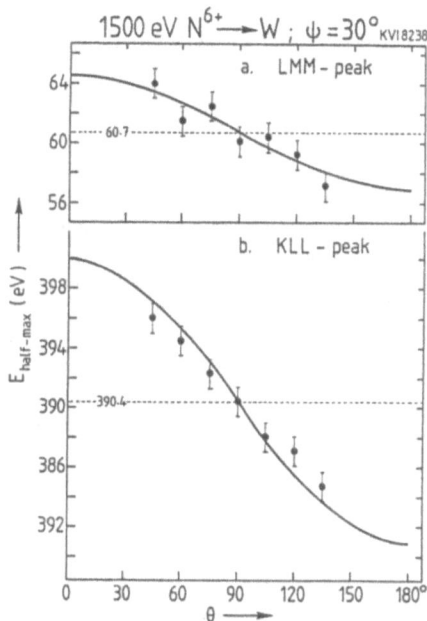

Figure 4: Doppler shift of LMM- and KLL-peaks due to N^{6+} collisions on tungsten at 1500 eV and an angle of incidence of $\Psi = 30$ deg. The lines represent calculated shifts, assuming electron emission from the projectile on the incoming part of the trajectory.

the projectiles which in a preceding process have captured electrons from the surface. A more direct proof that these electron are in fact emitted from the projectile is based on the Doppler shift, which is observed when the electrons are detected at different angles with respect to the ion beam direction. The result of such a measurement is shown in fig. 4. The LMM peaks as well as the KLL peaks are shifted to higher energies when the electron detector is positioned such that the projectiles move towards the detector ($\Theta < 90$ deg), and to lower energies when the detection takes place at angles such that the projectiles move away from the detector ($\Theta > 90$ deg). Quantitative agreement between measured and calculated Doppler shifts is obtained if the calculations are based on the assumption that electron emission takes place on the incoming part of the trajectory before the projectile actually hits the surface.

In the first place one might be tempted to believe that the LMM- and the KLL-transitions are two steps of a whole cascade in which electrons from the valence band of the metal are captured quasi resonantly into high levels of the projectile and then cascade down to the lower levels. This is in fact the mechanism which was suggested long ago by Arifov et al [6]. However, a more detailed analysis of the precise energetic position of the Auger lines, their relative intensity and their different behavior as function of the collision energy indicates, that this is not the case. And it is just the deviation from the initially expected behavior, that gives us more insight into the history of the collision process.

4.1 The energetic position of KLL and LMM Auger lines

It is worthwhile to study the energetic position of the various Auger lines in some more detail because this position is influenced by the specific charge state of the projectile during the emission process. Fig. 6. shows the KLL-peak resulting from 150 eV impact of hydrogen like O^{7+} ions together with an indication of the energies of Auger lines resulting from singly K-shell ionized oxygen. The lines observed by us are clearly shifted to somewhat higher energies. This can be taken as an indication [7] that the KLL transitions take place in a completely neutralized projectile, having one K-shell electron and seven L-shell electrons. We have estimated the energies of the corresponding KLL Auger energies and these are indicated in the figure.

A similar analysis can be done for the LMM Auger peak, which is shown in more detail in fig. 5. For an estimate of the line energies we have assumed that the transitions take place in a completely neutralized projectile, having one K-shell vacancy and the remaining seven electrons distributed in different ways over the L- and M-shell. Our estimates indicate that most transitions take place when the L-shell is filled with one or two electrons. From this we can conclude that filling of the L-shell with electrons proceeds via LMM transitions only in the first stage.

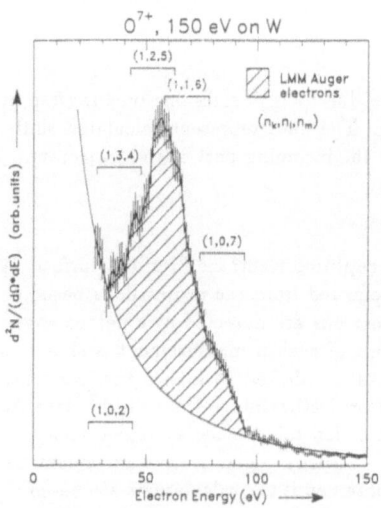

Figure 5: LMM Auger lines resulting from 150 eV collisions of hydrogen like O^{7+} ions on tungsten. Indicated are estimated positions of Auger lines resulting from transitions with different numbers (n_K, n_L, n_M) of electrons in the K- L-and M-shell respectively.

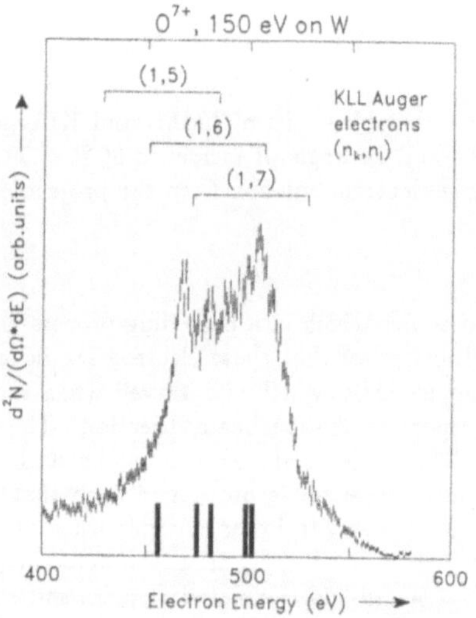

Figure 6: KLL Auger lines resulting from 150 eV collisions of hydrogen like O^{7+} ions on tungsten. The bars indicate the energetic positions of well known Auger lines in singly K-shell ionized O.

4.2 Relative intensities of LMM and KLL Auger lines

Another indication, that the L-shell is only partly filled by LMM transitions is based on an analysis of the relative line intensities. In hydrogen like projectiles only one KLL transition can take place, but for $Z < 10$ a total of $Z - 2$ LMM transitions are necessary,

Table 1: Relative intensities of the LMM and KLL electron peaks intergrated over the given energy ranges.

X^{q+}	E_{prim} (eV)	LMM range (E_{max}) (eV)	KLL range (eV)	$\gamma_{LMM}/\gamma_{KLL}$
C^{5+}	150	<25–56 (33)	200–300	>0.9
N^{6+}	90	26–70 (44)	315–425	1.3
O^{7+}	150	33–95 (60)	420–570	2.0
Ne^{9+}	600	60–150(89)	700–900	1.6

if the L-shell is to be filled exclusively by LMM transitions. This would result in a ratio of $Z - 2$ for the relative line intensities. We have investigated collisions of various hydrogen like projectiles on tungsten and compared the corresponding line intensities.The results are given in table 1, and it can be seen that the ratio γ is much smaller than $Z - 2$, in most cases even below $\gamma = 2$. This implies that a side feeding mechanism different from Auger cascades has to contribute appreciably to the filling of the L-shell [8].

4.3   Collision energy dependence of the electron intensities

Still another indication for the side feeding process mentioned above comes from an analysis of the different behavior of LMM and KLL peaks as a function of the collision energy. In fig. 7 we have plotted the peak intensities as a function of $1/v$, with v being the velocity of the projectile. The x-axis can as well be regarded to represent the time which is available for capture and emission of electrons in front of the surface. At small values of $1/v$, i.e. when the time in front of the surface apparently becomes too short for capture and emission of electrons, the peak intensities decrease. Remarkably enough however the

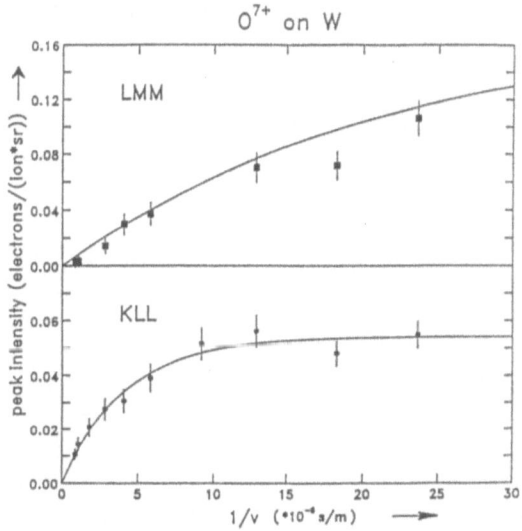

Figure 7: Intensities of the LMM and the KLL peak resulting from O^{7+} collisions on tungsten as a function of the inverse collision velocity. Main feature is the fact that the LMM peak suffers more from a lack of time in front of the surface than the KLL peak.

LMM peaks exhibit a more pronounced decrease and obviously suffer more from a lack of time than the KLL peaks. One can even deduce typical velocities or times from the energy dependent intensity decrease. If we assume an exponential decay in time described by $I = I_0 exp(t/\tau) = I_0 exp(R/v\tau)$, with R being a characteristic distance where the decay starts, we find values of $\tau(LMM) = 2.5 \cdot 10^{-14}$ s and $\tau(KLL) = 4.6 \cdot 10^{-15}$ s respectively, where we have estimated R from the relation $R = 2q + 7$ given by Snowdon [9]. Clearly the KLL decay takes place faster than the LMM decay and therefore can not exclusively be dependent on the latter one as a feeding process.

5 Models for the neutralization process

The various dynamic processes taking place during the approach of the multiply charged ion to the surface can best be discussed in terms of a potential diagram as depicted in fig. 8. With decreasing distance R the potential barrier between the ion and the solid is lowered such that electrons from the conduction band can pass over to the projectile. This leads to a resonant neutralization (process RN). Snowdon [9] has estimated that this process starts at R-values connected to the projectile charge q by $R = 2q + 7$. In subsequent steps electrons can be emitted in Auger ionization processes (AI). Competing processes for a more direct filling of lower levels are the Auger capture (AC), in which two electrons from the conduction band are involved, and the quasi resonant transfer (QT) of electrons, where electrons from deeper levels in the target atoms are involved. The Doppler shifts of emitted electrons discussed above indicate, that Auger capture does not play an important role in the cases studied by us. In the following we will therefore consider the competition between resonant neutralization with subsequent Auger cascades on the one hand and quasi resonant transfer on the other hand for filling the inner shells. For the L-shell filling in the hydrogen like projectiles we can rather directly estimate the contributions from these two processes, based on the experimental observations discussed above: (i) the ratio of LMM and KLL intensities and (ii) the exact line positions, re-

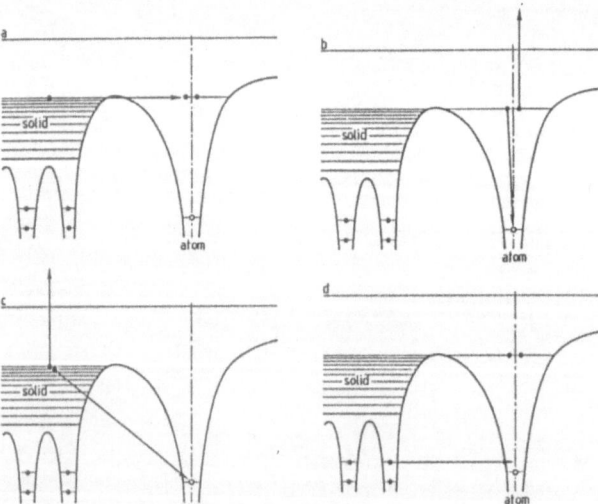

Figure 8: Mechanisms for electron capture and emission in a schematic potential diagram: a. resonant neutralization b. autoionization c. Auger capture d. quasi resonant neutralization

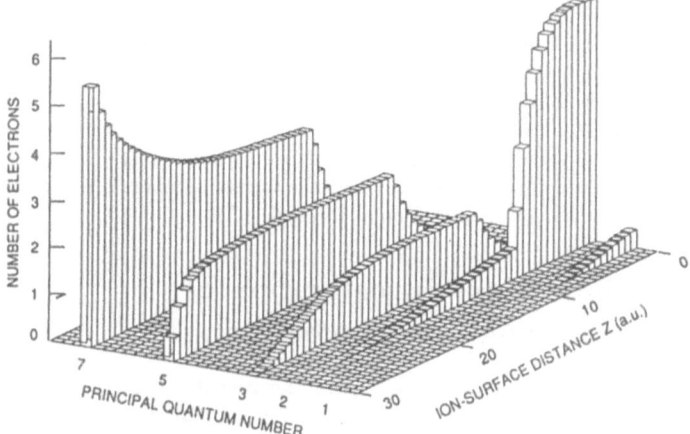

Figure 9: Numerical simulation of level populations resulting from 24 keV collisions of N^{6+} on a metal surface at an angle of incidence of 10 deg as a function of projectile-surface distance, from ref. [10].

flecting the average number of L-electrons present in the L-shell during LMM and KLL transitions.

In O^{7+} collisions on tungsten at 150 eV for instance we arrive at the result that on the average roughly 2 electrons have entered the L-shell via LMM transitions, whereas eventually 6 electrons will have entered via quasi resonant transfer. Similar numbers apply for the other hydrogen like projectiles.

Next one may of course ask whether the M-shell is mainly filled via Auger cascades from higher lying projectile levels or if also here side feeding via quasi resonant transfer yields an important contribution. At this moment this question can not be answered on the basis of experimental results. However from theory there are indications that in fact Auger cascades will be dominantly responsible for filling of the L-shell. First of all Zeijlmans van Emmichoven et al [10] performed numerical simulations of resonant capture followed by Auger cascades for N^{6+} ions approaching a metal surface, in which they calculated numerically populations of various levels as a function of the distance between projectile and surface. The result of their final simulation is shown in fig. 9. In order to obtain agreement with the experimentally determined yield of KLL electrons they had to make two assumptions: (i) they had to decrease the calculated Auger lifetimes by a factor of 20, and (ii) they had to assume a side feeding process for additional population of the L-shell. In view of our experimental results discussed above the second assumption can certainly be regarded as justified. Regarding the first assumption it is interesting to look at the results of Hansen and Vaeck [11], who have recently calculated Auger transition rates of multiply excited ions and atoms. For hydrogen like N^{6+} ions resonantly neutralized by capture into $N(1s7p^6)$ they find a lifetime of only $3.5 \cdot 10^{-16}$ s [11], whereas Zeijlmans van Emmichoven et al in their simulations had in the first place used a lifetime of $2 \cdot 10^{-15}$ s [10]. The results of Vaeck and Hansen therefore imply that also the first assumption of Zeijlmans van Emmichoven et al is justified. We therefore conclude that the low energy electrons including those from LMM transitions are due to resonant neutralization of the projectile followed by Auger cascades, whereas the filling of the L-shell and the subsequent emission of KLL electrons is to a large extent caused by quasi resonant electron transfer from lower lying levels.

Several questions are still open. It is e.g. by no means clear at which distance from the surface the sidefeeding takes place, and possibly a considerable fraction of the KLL electrons is emitted at a time when the projectile is very close to or even has penetrated into the surface. If this sidefeeding is due to a direct electron capture from inner atomic target orbitals one would expect also target Auger electrons to be ejected. However these have been observed only with very small intensities [1]. It would be interesting to know more details about the processes at small distances from the surface because these are the most likely candidates for inducing a modification of the surface.

6 Tentative application: investigation of magnetic short range surface ordering?

It might be interesting to speculate about more applied experiments, and what can be expected on the basis of the knowledge obtained from the study of multiply charged ion-surface collisions. One experiment planned in our group involves magnetized surfaces. The idea for this is based on the possibility, that preferentially electrons with parallel spins are transferred from the surface to the projectile ion, if capture takes place out of one magnetic domain. Rau and Eichner have performed related experiments in which two electrons were captured from a Ni surface [12] by D^+ to form D^- in a triplet state. From experiments with gaseous targets it is in fact well known that during the electron transfer process the spin is conserved. Also during Auger cascades spin flips are unlikely. One therefore expects neutralized projectiles in extremely high spin states. This should be reflected in the energy spectra of ejected electrons and give e.g. rise to so called hypersatellite structures. The relative intensity of such hypersatellites might be used to extract information about the magnetic surface ordering in the surface area from which the electrons were captured. This area can be varied widely by using different angles of incidence for the projectiles and different kinetic energies, and in this way one could possibly obtain information about the short range magnetic surface ordering, which is difficult to obtain in another way. There are in fact indications that the short range magnetic order is conserved at elevated temperatures even above the Curie temperature.

In view of the conclusions drawn in the last chapter one should especially look for the low energy part of the electron spectra to observe such spin effects, since the high energy electrons are mainly produced via processes involving capture from lower levels, for which spin effects are unlikely to play an important role.

References

[1] S. T. de Zwart, A. G. Drentje, A. L. Boers, and R. Morgenstern, *Surf. Sci.*, **217** (1989) 298.

[2] R. Geller and B. Jacquot, *Nucl. Instr. Methods*, **202** (1982) 399.

[3] A. G. Drentje, *Nucl. Instrum. Methods Phys. Res. B*, **9** (1985) 526.

[4] S. T. de Zwart, T. Fried, D. O. Boerma, R. Hoekstra, A. G. Drentje and A. L. Boers, *Surf. Sci.*, **177** (1986) L939.

[5] J. P. Briand, L. de Billy, P. Charles, S. Essabaa, P. Briand, R. Geller, J. P. Desclaux, S. Bliman, and C. Ristori, *Phys. Rev. Lett.*, **65** (1990) 159.

[6] U. A. Arifov, L. M. Kishinevskii, É. S. Mukhamadiev, and É. S. Parilis, *Sovjet Phys.-Tech. Phys.*, **18** (1973) 118.

[7] L. Folkerts and R. Morgenstern, *J. Physique C1*, **50** (1989) 541.

[8] L. Folkerts and R. Morgenstern, *Europhys. Lett.*, **13** (1990) 377.

[9] K. J. Snowdon, *Nucl. Instr. Methods Phys. Res. B*, **34** (1988) 309.

[10] P. A. Zeijlmans van Emmichoven, C. C. Havener, and F. W. Meyer, *Phys. Rev. A*, **43** (1991) 1405.

[11] J. E. Hansen and N. Vaeck, *private communication.*

[12] C. Rau and S. Eichner, *Phys. Rev. Lett.*, **47** (1981) 939.

Semiconductor Detectors for Nuclear Radiation – A Review

S. Kalbitzer

Max-Planck-Institut für Kernphysik, Postfach 10 39 80,
W-6900 Heidelberg, Fed. Rep. of Germany

A historical review of the development of semiconductor radiation detectors is given with focus on the major breakthroughs in the areas of material development, processing techniques, device design and electronic equipment.

Introduction

Approximately 40 years ago, on September 29, 1949, a Physical Review Letter was published in which K.G.McKay /1/ described a barrier layer detector recording α-particles. According to the state of semiconductor technology at that time the single crystal was Ge. The sensitive region around a bronze point contact was estimated to have a diameter of 10-100 μm. At an operating voltage of 10 V the signal-to-noise (rms) ratio was about 200, i.e. a 5 MeV signal had an energy width of about 60 keV (FWHM).

It did not take more than one decade until nuclear physics labs all over the world selected one of their students to make these astonishingly simple solid ionization chambers. All what seemed to be needed was a rod of single crystal semiconductor material commercially available, some cheap chemical etchants and an evaporation apparatus, which was anyhow available at the target lab.

At that time, semiconductor physics was considered some kind of alchemy and, indeed, many adventurous recipes of how to make good detectors were circulating on the black market - and many sneering remarks were to be heard when the detectors failed in the hands of the users.

The remarkable state of semiconductor detector technology today has been the result of continued efforts in four major areas:

1. semiconductor material,
2. processing techniques,
3. device design,
4. electronic equipment.

In the following brief overview over the achievements in this field we shall emphasize the breakthroughs in the different areas and, consequently, only refer to the first publications in each case.

Detector history

In Table 1 we have listed the main historical events in the semiconductor detector field during the last four decades.
After the demonstration of the detector principle by recording α-particles with a Ge point contact rectifier in 1949 more suitable detector structures were investigated. Purdue University [2] and Bell Labs [3] reported the successful use of p-n junction devices consisting of a Ge crystal grown with changing doping (Fig.1). The p-n junction, superior to the point contact within many respects, allowed for the extraction of basic physical data from the measurements, e.g. the mean energy for electron-hole pair production was quoted to be w = 3.0±0.4 eV. Today's accepted mean value is w = 2.96 eV in Ge at 77 K.
These numbers were recognized as very promising as regards ionisation statistics: a 3 MeV particle will produce about 10^6 electron-hole pairs, so that a statistical fluctuation of 0.1% in the energy signal will be obtained (for zero correlation or a Fano factor of f = 1). This was much better than any other known system, e.g. gas counters with w $\sim$ 30 eV and scintillation-photomultiplier systems with w $\sim$ 300 eV.

Table 1: Milestones in the development of semiconductor detector systems for nuclear spectrometry

1949	Ge point contact diode	[1]
1950/1	Ge pn junction diode	[2,3]
1959	Au/Si surface barrier diode	[4,5]
1962	1D-position sensing Au/Si diode with resistive back layer	[6]
1962	Si(Li) drifted diode	[7]
1962	Ge(Li) drifted diode	[8]
1964/6	ion implantation studies in Si	[9,10]
1965	Si X-ray diodes and FET preamplifier	[16]
1965	Ge X-ray diodes and FET preamplifier	[17]
1967	ion implanted Si diode	[11]
1967	ion implanted 1D and 2D position sensing Si diodes	[12]
1969	ion implanted buried-layers in Si	[18]
1970	hyperpure Ge	[14]
1972	ion implanted Ge diode	[13]
1973	ion implanted $\Delta E/E$ Si detector telescope	[19]
1980	planar technology for Si detectors	[20]
1987-	various Si devices: drift chambers, variable telescopes, low-capacitance and current diodes, strip detectors, integrated electronics	[22-26]

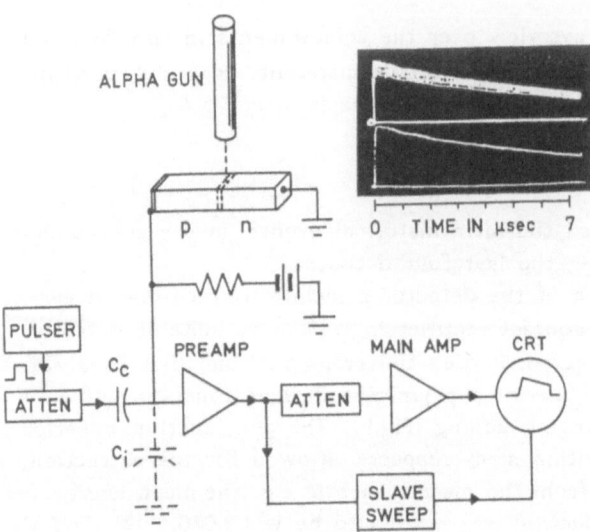

Fig.1: The experimental set-up by McKay /3/ for the detection of Po-α particles with a p-n junction grown into the Ge single crystal; some pulses are shown in the inset. Note the short pulse rise-time and the good signal-to-noise ratio of about 1% of the ionisation induced signal.

Diamond counters did not receive much attention because of the lack of suitable crystals.

With the advent of single-crystal Si a dramatic increase of the interest in solid state particle detectors took place, especially when the surface barrier detector was introduced about 1960 /4-5/. Although easy to fabricate and operable at room temperature the handling was delicate. A single finger print sufficed to damage the 100 nm Au layer and render the detector unusable. Nevertheless, a widespread application was seen and crystal growers were pressed to produce ultrapure Si with near-intrinsic resistivities of about 100 kΩ-cm in order to satisfy the demand for sensitive thicknesses of several mm. Zone refining techniques, however, turned out to be not efficient enough for removing the natural B contamination from the ingots, a direct consequence of the fact that the segregation coefficient of B in Si is very unfavorable, $k \sim 0.8$. Dozens of cycles would be needed to reach the near-intrinsic concentration level of some 10^{10} carriers/cm^3.

In 1962, the principle of a position sensing detector was realized /6/ by using Si surface barrier techniques with an evaporated resistive metal layer serving as a charge-dividing back electrode. Despite many efforts, however, the problems of ageing of the metal film and of matching its resistance to the optimum RC-value were not solved.

The application of the Li-drift technique overtook the efforts of producing thick detectors from multiturn zone-refined near-intrinsic Si /7/.
Its unique ability to exactly compensate any local acceptor density - in
contrast to the neutron compensation technique where a selected average compensation level of an acceptor background can be achieved by
transmutation of Si into P atoms - rendered it a world-wide used
tool, although the fabrication process of several mm thick Si detectors
took weeks time. When used in actual accelerator experiments for
nuclear particle detection, such as p,d,α etc., however, their sensitivity
towards radiation damage put up severe restrictions as regards their
life time. Since the effect of γ-rays in defect production is many orders of magnitude smaller, the application of Ge(Li) detectors /8/ for γ-
ray spectroscopy continued and the main failure of these devices, permanently to be stored at the temperature of liquid nitrogen, was caused
by accidental warm-ups of the cryostats.

Within this same period of 1960-1970 other important developments
took place in materials processing techniques and electronic pulse processing.

There had been some attempts of making Si p-n junction detectors
by ion implantation /9,10/. Dopant activation and rectifying properties
were observed after thermal treatments, but the reverse currents were
very much higher than those of surface barrier detectors, typically a-
bout 1 $\mu A/cm^2$ equivalent to a noise level of about 30 keV(FWHM).
In addition, implantation energies of 100 keV had been used so that
projected ranges of about 0.4 μm for B and considerably thicker surface
dead layers were to be expected in an operating counter.

In 1967 the first high-quality p-n junction Si detector was made by ion
implantation, with B and P ions of 2-5 keV energy for the front and
back contact, respectively /11 /. After annealing at about 400°C the reverse current was 1 μA leading to a line width of about 20 keV(FWHM)
similar to values obtained with surface barrier Si detectors (Fig.2).

Further analysis led to the result that surface currents rather than bulk
sources gave rise to this excess current, very much as in surface barrier detectors. From the known minority carrier life time of the order
of 1 ms one would deduce bulk generation currents lower by some orders of magnitude. The α-lines in Fig.2 are highly symmetric indicating
complete charge collection. The window thickness was measured to be
$w \sim 0.2$ μm by the detector-tilting technique; from the widths of the α-
peaks and the pulser line an upper limit for the window thickness of
$w \sim 0.25$ μm is derived by using the energy straggling relation $\Delta E (keV) =$
$25 [w(\mu m)]^{1/2}$.

In the same year, the ion implantation technique was successfully applied to the fabrication of Si position sensing detectors in which B and
P implants served to make the front p^+-n contact and the resistive n-n$^+$

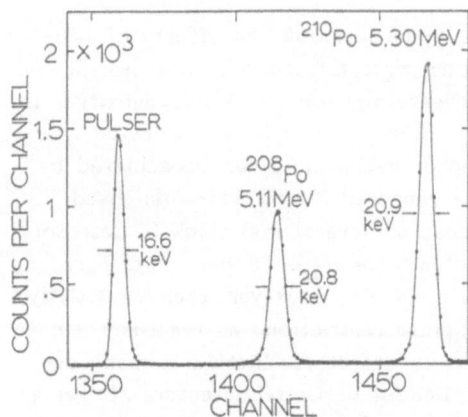

Fig.2: Ion-implanted p⁺-n-n⁺ Si detector; 2-5 keV 10^{14} B/cm² and 10^{14} P/cm², annealing temperature 400°C, window thickness 0.2 μm; bias 50 V, reverse current 0.5 μA at room temperature /11/. Note the absence of tailing in the α-peaks.

Fig.3: Two-dimensional position sensing Si detector made by ion implantation. Both top (p⁺) and back (n⁺) layer are precision resistors for charge division. The recorded pattern is due to Am-α particles through a mask with the letters XY as apertures /12/.

back layer, respectively /12/. Due to the inherent precision of the ion implantation process the electrical parameters of the charge dividing resistor can be tailored to the order of 1%. Fig.3 shows the spectrum of the first two dimensional position sensing detector realized by ion implantation of high resistivity Si.

These results were obtained at a time when there was the general belief that the severe and complex radiation damage by heavy ions in semicon-

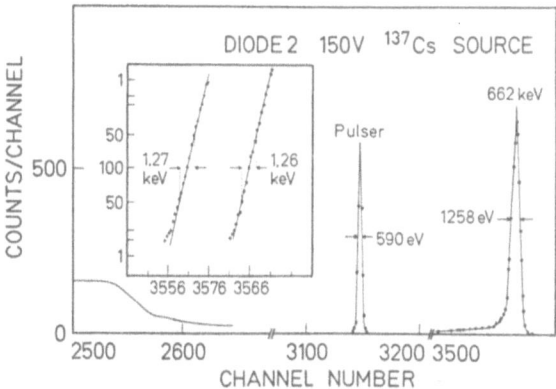

Fig.4: Ion implanted hyperpure Ge detector with high energy resolution; the Gaussian plot (inset) demonstrates both detector and process quality /13/. From the energy width of the γ-ray peak a Fano factor of about 0.1 is derived.

ductors could not be repaired by annealing at practical temperatures, at least not to the necessary extent that charge transfer over typical distances would be sufficient, i.e. that especially minority carrier life times could be restored to about the 1 μs level.

Therefore, the above results obtained on semiconductor particle detectors with transport distances of up to 1 mm had by far greater significance than just in the small, special field of Si radiation detectors. It became immediately clear to semiconductor experts that all other semiconductor devices based on charge transfer over much smaller distances, of the order of 1 μm, should also be producible in this way.

The first two international conferences on ion implantation in semiconductors, at Thousand Oaks (1970) and Garmisch-Partenkirchen (1971), corroborated these results and added other detector types, e.g. high purity Ge p-n junction detectors /13/, to the list of success. Fig.4 is a spectrum of the γ-ray line at 0.662 MeV of a ^{137}Cs source, where the high quality of the Ge crystal is demonstrated by the Gaussian plot. This material had just become available after a remarkable development started some years before: the growth of hyperpure Ge with net dopant levels of as little as 10^{10} electrical centers per cm^3 /14/, (Fig.5). Contrary to the case of Si, the segregation coeffients of all relevant dopants are about k ~ 0.1 rendering the respective purification techniques quite effective. These high-purity Ge detectors can be stored at room temperature without deterioration as there is no highly mobile doping impurity, such as Li. Only for optimum operation conditions low temperatures have to be applied. This minor inconvenience, quite tolerable in a laboratory environment, will become a major one as soon as other requirements have to be met, e.g. portability for use outside

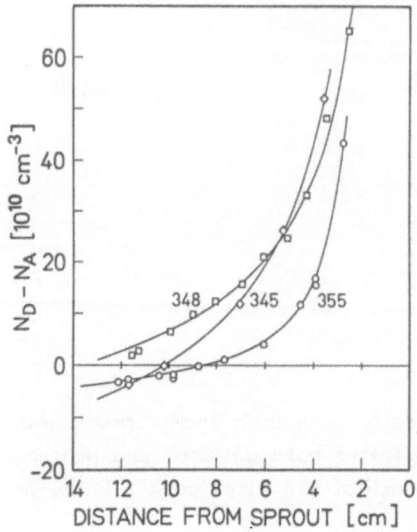

Fig.5: Impurity profiles, i.e. the net electrical carrier density distribution as the difference between donor and acceptor concentration, $N_D - N_A$, as a function of the position of the electrical probe along the hyperpure Ge single crystal ingot.

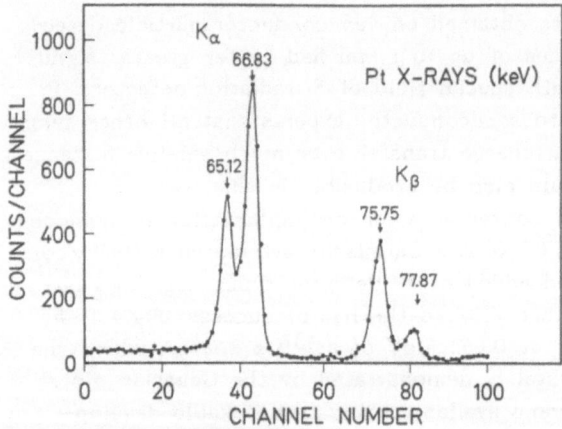

Fig.6: High resolution X-ray spectrometer consisting of a Si(Li) detector and a field-effect transistor preamplifier /17/ operated at 77K. The K_α and K_β X-rays of Pt are resolved with a precision of 1.0 keV(FWHM); the corresponding pulser width (not shown) is 0.7 keV.

the lab. It will lead to far here to review the numerous attempts to overcome the basic physical limitations of semiconductor materials with (indirect) band-gaps of about 1 eV by developing detectors from materials with band-gaps of at least 1.5 eV, in order to reduce the reverse currents to insignificant levels, and high atomic number for better detection properties, especially for γ-rays /15/.

Within this same decade, the field-effect transistor was introduced and yielded hitherto unknown low noise levels. The best amplifier tubes at that time had noise levels of about 30 keV (FWHM), whereas now figures of about 1 keV at 300 K and later on of about 100 eV at the temperature of liquid nitrogen could be achieved. This again made the Si detector X-ray analysis possible which has become a widespread technique for chemical analysis /16,17/ (Fig. 6).

A further successful application of the ion implantation technique, recently receiving considerable attention for future high integration processes, was to produce buried layers for ΔE-E-detector telescopes /18, 19/. To this end, MeV B ions have to be implanted to form the deep p^+-layer serving as a common contact to both diodes. In practice, according to the requirements of the actual experiment, a certain optimized thickness of the energy-loss section should be provided. Thus, a major inconvenience arises from the necessity of having a complete set of different telescopes available. The true solution to this problem would be to have an integrated telescope system with a variable energy-loss section. This goal has indeed be realized by novel material processing techniques and ingenious detector design, as we shall describe in more detail below.

The next important step, at the end of this decade was the introduction of the planar technology to detector fabrication /20/. As had been known for the whole detector history, the current levels of the order of $1\mu A/cm^2$ in Si surface barrier detectors, with room temperature fabrication steps involved only, clearly indicated excess currents from surface sources. Guard rings, epoxy resins and the like were used to suppress them- with moderate success only. Thus, it was logical to protect the sensitive part of the surface region by adopting the planar technology used in Si device fabrication by industry in which SiO_2 is used for passivation (Fig. 7). This measure brought an enormous success: currents of formerly 1 $\mu A/cm^2$ dropped to 1 nA/cm^2, the contributions to detector noise from 30 to 1 keV (FWHM). Now other sources, such as window thickness effects, became dominant in the resolution of ionizing nuclear particles, e.g. of about 10 keV for 5 MeV α particles through a dead layer of 0.2 µm Si. As a concomitant disadvantage of the detector current reduction may be considered that the equivalent radiation damage level - i.e. the irradiation dose needed to rise the initial reverse current to twice its value - is lowered by the same factor.

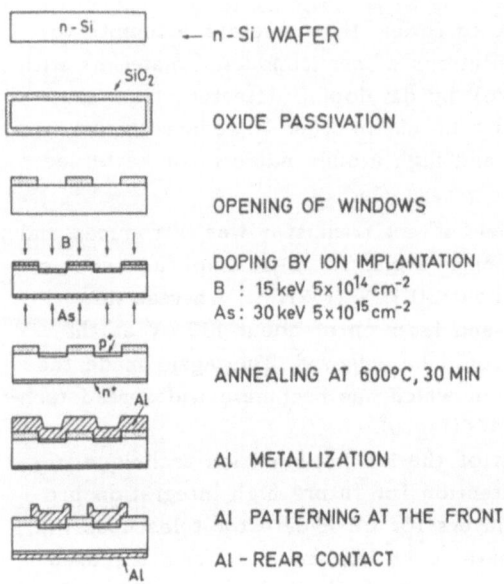

n - Si   ← n - Si WAFER

SiO₂   OXIDE PASSIVATION

OPENING OF WINDOWS

B   DOPING BY ION IMPLANTATION
B : 15 keV 5 × 10¹⁴ cm⁻²
As: 30 keV 5 × 10¹⁵ cm⁻²

ANNEALING AT 600°C, 30 MIN

Al METALLIZATION

Al PATTERNING AT THE FRONT
Al - REAR CONTACT

Fig.7: Si detector fabrication steps by using the planar Si device technology /20/ based on addition and removal of thermally grown
SiO₂ protective layers. Note that the annealing step of 600°C
is a compromise between life-time conservation/impurity in-dif-
fusion and radiation damage removal/dopant activation.

A recent example of detector development using the same concepts
is a 2-ring, 24-segment annular Si detector for use in high-energy backscattering analysis and trace element measurements /21 / (Fig.8).
During the present decade, by combining the powerful tools of ion implantation and planar Si technology, novel and sophisticated detector
systems have been developped /22-25/: drift chambers, $\Delta E/E$- detector
telescopes with variable energy-loss section, low-capacitance diodes,
charge coupled devices, strip detectors and integrated preamplifiers. Fig.9
shows the basic principle of a low-capacitance Si drift chamber for use
as a $\Delta E/E$-telescope with adjustable thickness of the ΔE section /23/.
Fig.10 demonstrates the complexity of advanced detector systems by
the example of of a Si drift chamber. Here a 300 μm thick Si wafer
was processed by using the above explained techniques of photolithography and ion implantation in order to obtain a large number of electrodes at a 150 μm spacing /23/. Si micro-strip detectors, with position
resolutions down to the 10 μm level, are nowadays met in many high-
energy physics experiments.
Further progress has been achieved by integrating the preamplifiers, or
at least the most important parts thereof, of the pulse-processing electronics onto the same Si wafer as the particle detector by using special

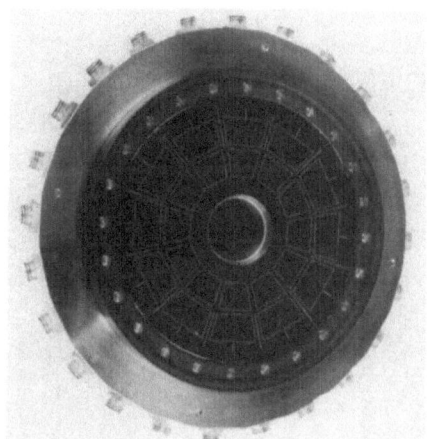

Fig.8: Photograph of a 2-ring, 24-segment annular Si detector fabricated by the planar technique /21/. The connection of the individual detector areas to the outer microdot connectors is through
Al wires bonded to Au plated pads in a circular arrangement at
an intermediate position. The whole set is designed to be heated
up to a minimum/maximum temperature of 200/400°C.

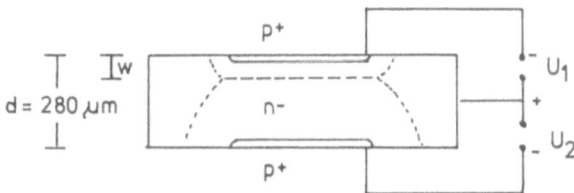

Fig.9: Scheme of a Si drift chamber to be used as a ΔE/E telescope
with a variable energy-loss section w. For minority carriers
(holes) the bulk is divided into two independent detectors due
to the shape of the electrical field inside the depleted zones.

circuit components, e.g. MOS or junction FET's, compatible with the low-
temperature detector fabrication process (Fig.11). In this way, as Fig.12
demonstrates, X-ray resolutions of about 130 eV (FWHM) have been obtained for 6 keV X-rays with the detector/FET unit at 130 K /26/.
Due to the reduction of both input capacitances and current sources to
the extremely low levels of about $C \sim 0.1$ pF and $j \sim 1$ pA a pulser
width of only 45 eV (FWHM) was measured. This figure corresponds to
a statistical fluctuation of about 5 electron-hole pairs. Although this figure may still be cut to half, the limiting factor is already the ionisation
statistics as is readily inferred from the above X-ray line-widths. A Fa-

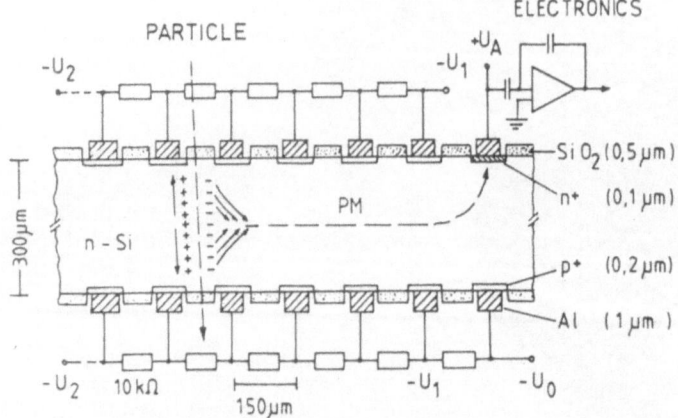

Fig.10: Silicon drift chamber with p^+-electrode arrays on both sides of the wafer. A transverse drift field is generated by the voltage-dividing resistor chain. The carrier flow due to an ionizing event is indicated: the holes drift in vertical direction towards the nearest p^+ electrodes, whereas the electrons move laterally along the potential minimum towards the outer n^+ electrode.

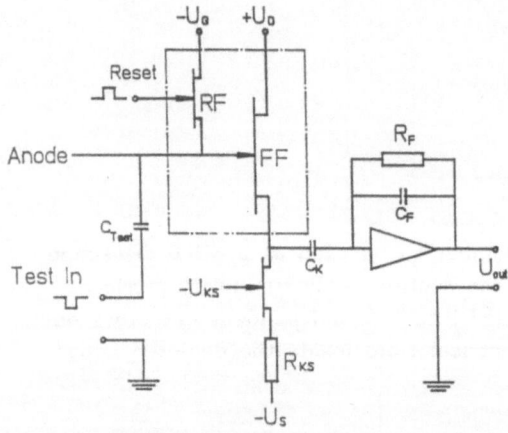

Fig.11: Pulse-processing electronics for a charge-coupled Si detector device; the input and reset field-effect transistors, FF and RF, are integrated onto the Si detector chip.

no factor of 0.12 indicates the preservation of the high quality of the detector grade Si material throughout the various processing steps. These developments indicate that future devices will become even more complex and powerful. The trends can clearly be seen to integrate, to a steadily growing extent, the analog electronics onto the same Si wafer on which the various detectors are produced, gaining compactness,

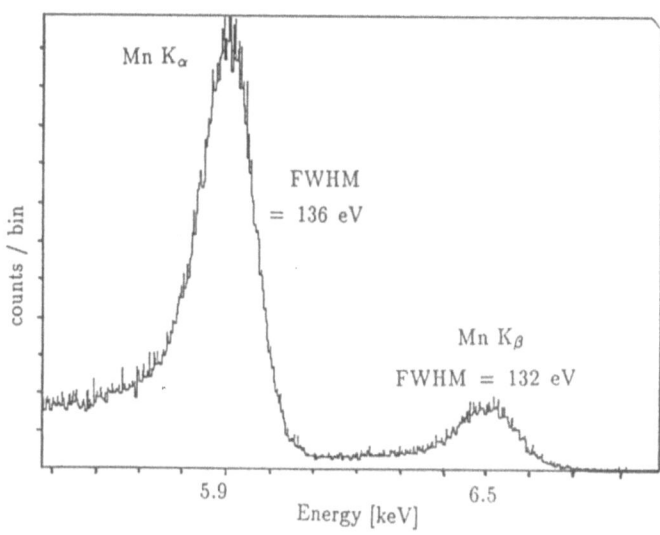

Fig.12: X-ray spectrum of Mn-K$_\alpha$ and Mn-K$_\beta$ lines as recorded with a charge-coupled Si detector device and integrated preamplifier input stage after Fig.11 at 130 K. The pulser width (not shown) is 45 eV (FWHM).

resolution and speed. The overnext step may be to include also further integration measures for digital signal processing.

The developments in the detector materials area have been less exciting. Nevertheless, mention must be made of a recent attempt to use a new class of detector materials: amorphous silicon of the glow-discharge type, a-Si:H$_x$ with x ~ 5 at% H, for large area devices. Although some encouraging results have been reported on small area detectors with some µm thick layers /27/, the inherent charge collection problems of this type of material are well known: both mobility µ and minority carrier life-time τ are very much smaller than in the crystalline phase, e.g. µτ ~ $10^{-7}/10^{-8}$ cm^2/V for electrons/holes in a-Si:H, whereas µτ ~ 1 cm^2/V in c-Si (Table 2). The corresponding drift length ("Schubweg"), λ = µτE, in a-Si:H diodes operated at electrical field strengths of E ~ 10 kV/cm would be of the order of 10^{-3}-10^{-4}cm. This value is not sufficient for high energy-resolution work requiring λ/d ~ 100 - 1000 for charge collection efficiencies of about 1% and 0.1%, respectively, where d is the thickness of the depletion layer of the detector; for c-Si we have λ ~ 10^4 cm! Also the timing properties are insufficient for many purposes: for a 10 µm thick a-Si:H diode with a maximum electron drift velocity of about v ~ 10^{-1} x 10^4 = 10^3 cm/s the charge collection time would amount to t ~ 1µs. This has to be compared with (saturated) drift velocities of 10^7cm/s in c-Si allowing for subnanosecond timing.

Table 2:   Basic transport parameters of c-Si and a-Si:H

	c – Si	a – Si:H
μ_e (cm^2/Vs)	1350	1
μ_h (cm^2/Vs)	480	10^{-2}
τ_e (s)	10^{-3}	10^{-7}
τ_h (s)	10^{-3}	10^{-6}
$\mu_e\tau_e$ (cm^2/s)	1.35	10^{-7}
$\mu_h\tau_h$ (cm^2/s)	0.48	10^{-8}
λ_e (cm)$|_{10kV/cm}$	10^4	10^{-3}
λ_h (cm)$|_{10kV/cm}$	5×10^3	10^{-4}

In addition, material degradation of a-Si:H has been observed under intense photon irradiation. It remains to be seen whether this material in
form of about 10 μm thin films can be used as a PIXEL particle detector,
i.e. not as a device with a high energy resolution, but as a monitor device, with a total sensitive area of the 1 m^2 size.
What will the developments of the near future be? I would expect to
see even more refined designs of semiconductor detector systems on Si
basis rather than the development of new detector materials excelling
the performance of c-Si or c-Ge. Of course, one may wish to have SiC
systems for higher ambient temperatures or for radiation environments,
or a high Z material, such as CdTe or HgI$_2$, for even better γ-ray detection, and so on. But, as history tells, material development is a difficult and hence a slow process, even if one would know what the suitable elemental or, less likely, compound semiconductor with an indirect
band-gap, good electron-hole mobilities, etc. would be. Thus, progress
in system complexity, using the well-established processes of ion implantation and photolithography, on the basis of an existing top quality
semiconductor material such as Si, is the likely development of the coming years. A further step in the processing development could be the
elimination of chemistry by the direct writing of the detector and electronic structures with ion micro-beams. Table 3 summarizes the general
developments in the semiconductor detector history.
We all are eager to see what resumé will be drawn on the next major
birth day of the charge collection type of semiconductor detector, which
will be the fiftieth at the end of this millenary.

Table 3: General developments of semiconductor detector systems for nuclear spectrometry

	1950	1960	1970	1980	1990	2000
MATERIAL	Ge (C)	Si Ge	Si Ge	Si Ge	Si Ge	
PROCESS	POINT CONTACT P-N JUNCTION	SCHOTTKY Li DRIFT IMPLANTATION	IMPLANTATION SCHOTTKY PLANARTECH	IMPLANTATION PLANARTECH SCHOTTKY	IMPLANTATION PLANARTECH	
DETECTOR	ENERGY	ENERGY ENERGY LOSS POSITION	ENERGY ENERGY LOSS POSITION	ENERGY ENERGY LOSS POSITION MULTI-FUNCTION	MULTI-FUNCTION COMPLEXITY	
ELECTRONIC	TUBES	FET	FET	FET INTEGRATION	HIGH LEVEL INTEGRATION	

References

/1/ K.G.McKay, Phys. Rev. **76** (1949) 1537

/2/ C.Orman, H.Y.Fan, G.J.Goldsmith and K.Lark-Horovitz, Phys. Rev, **78** (1950) 646

/3/ K.G.McKay, Phys. Rev. **84** (1951) 829

/4/ J.W.Mayer, J. Appl. Phys. **30** (1959) 1937

/5/ J.L.Blankenship and C.J.Borkowski, Proc. 6th Tripartite Instrumentation Conf., (1959) 75, and IRE Trans. Nucl. Sci. NS-7, No. 2-3, (1960) 190

/6/ K.H.Lauterjung, J.Pokar, B.Schimmer and R.Stäudner, Max-Planck-Institute Report 1962/V/10 (Heidelberg, 1962)

/7/ J.W.Mayer, IRE Trans. Nucl. Sci NS-9, No. 3 (1962) 124

/8/ D.V.Freck and J.Wakefield, Nature **193** (1962) 669

/9/ F.W.Martin, W.J.King and S.Harrison, IEEE Trans. Nucl. Sci. NS-**11** No.3 (1964) 280

/10/ F.W.Martin, S.Harrison and W.J.King, IEEE Trans. Nucl. Sci. NS-**13** No.1 (1966) 22

/11/ S.Kalbitzer, R.Bader, H.Herzer and K.Bethge, Z. Phys. **203** (1967) 117

/12/ S.Kalbitzer, R.Bader, W.Melzer and W.Stumpfi, Nucl. Instr. Meth. **54** (1967) 323

/13/ J.P.Ponpon, J.J.Grob, R.Stuck, P.Burger and P.Siffert, Proc. 2nd Int. Conf. Ion Implantation in Semiconductors, Garmisch 1972, I.Ruge and G.Keil, eds., Springer-Verlag, Heidelberg 1973, p.

/14/ R.D.Baertsch and R.N.Hall, IEEE Trans. Nucl. Sci. NS-**17** No.**3,** (1970) 235

/15/ J.W.Mayer, ref./29/ pp. 445

/16/ T.W.Nybakken and V.Vali, Nucl. Instr. Meth. **32** (1965) 121

/17/ E.Elad, Nucl. Instr. Meth. **37** (1965) 327

/18/ F.W.Martin, Nucl. Instr. Meth. **72** (1969) 223

/19/ A.Kostka and S.Kalbitzer, Appl. Phys. Lett. **23** (1973) 704

/20/ J.Kemmer, Nucl. Instr. Meth. **169** (1980)499

/21/ R.Günzler, V.Schüle, G.Seeliger, M.Weiser, K.Böhringer and S.Kalbitzer, Nucl. Instr. Meth. **B35** (1988) 522

/22/ E. Gatti and P. Rehak, Nucl. Instr. Meth. 225 (1984) 608

/23/ J.Kemmer and G.Lutz, Nucl. Instr. Meth. **A253** (1987) 365

/24/ V.Radeka, P.Rehak, S.Reccia, E.Gatti, A.Longoni, M.Sampietro, P. Holl, L.Strüder and J.Kemmer, IEEE Trans. Nucl. Sci. NS-**35** No.**1** (1988) 155

/25/ J.Kemmer, Nucl. Instr. Meth. B45 (1990) 247

/26/ H.Bräuninger, G.Lutz, N.Meidinger, P.Predehl, C.Reppin, W.Schreiber, L.Strüder, J.Trümper, E.Kenziora, G.Staubert, V.Radeka, P.Rehak, S. Rescia, E.Gatti, A,Longoni, M.Sampietro, P.Holl, J.Kemmer, U.Prechtel, H.Riedel and T.Ziemann, SPIE 1344 (1990) 404

/27/ V.Perez-Mendez, J.Morel, S.N.Kaplan and R.A.Street, Nucl. Instr.
Meth. B252 (1986) 478

For a general reading the following books may be consulted:
/28/ G.Dearnaley and D.C.Northrop, Semiconductor Counters for Nuclear Radiations, E. and F.N.Spon Limited, London, 2nd edition 1966
/29/ G.Bertolini and A.Coche, Semiconductor Detectors, North-Holland

The present state of art may be found described in the proceedings of
the Munich detector conference in 1989, published in Nucl.Instr.Meth.
A288 (1990).

Index of Contributors